THE PROBLEM OF SPACE TRAVEL
THE ROCKET MOTOR

Hermann Noordung (Herman Potočnik)

Edited by Ernst Stuhlinger and

J. D. Hunley with Jennifer Garland

Foreword by
Frederick I. Ordway III

Preface by
J. D. Hunley

The NASA History Series

NASA SP-4026

National Aeronautics and Space Administration
NASA History Office
Washington, D.C.   1995

Translation from German to English of "Das Problem der Befahrung des Weltraums, Der Raketen-Motor," Richard Carl Schmidt & Co.: Berlin, 1929, 188 pp., by:

SCITRAN
1482 East Valley Road
Santa Barbara, California 93108

Library of Congress Cataloging-in-Publication Data

Noordung, Hermann, 1892-1929.
    [Problem der Befahrung des Weltraums. English]
    The problem of space travel: the rocket motor/Hermann Noordung (Herman Potočnik); edited by Ernst Stuhlinger and J.D. Hunley with Jennifer Garland; foreword by Frederick I. Ordway III; preface by J.D. Hunley.
        p. cm. —(The NASA history series) (NASA SP; 4026)
    Includes bibliographical references and index.
    1. Space flight. 2. Space stations. 3. Astronautics.
I. Stuhlinger, Ernst, 1913- . II. Hunley, J.D., 1941-
III. Garland, Jennifer, 1968- . IV. Title. V. Series. VI. Series: NASA SP; 4026.
TL790.N613 1995
629.4—dc20
                                              94-36645
                                                CIP

Table of Contents

	Page
Foreword	ix
Preface	xv
Introduction	1
The Power of Gravity	3
The Practical Gravitational Boundary of the Earth	5
The Free Orbit	6
Maneuvering in the Gravitational Fields of Outer Space	8
The Armor Barrier of the Earth's Atmosphere	9
The Highest Altitudes Reached to Date	10
The Cannon Shot into Outer Space	11
The Reactive Force	12
The Reaction Vehicle	14
The Rocket	15
Previous Researchers Addressing the Problem of Space Flight	16
The Travel Velocity and the Efficiency of Rocket Vehicles	17
The Ascent	25
General Comments about the Structure of the Space Rocket	32
Proposals To Date	40
Comments Regarding Previous Design Recommendations	51
The Return to Earth	53
Hohmann's Landing Maneuver	55
Landing in a Forced Circular Motion	57
Landing in Braking Ellipses	59
Oberth's Landing Maneuver	61
The Result To Date	61
Two Other Important Questions	62
The Space Rocket in an Inclined Trajectory	62
The Space Rocket as an Airplane	65
The Space Station in Empty Space	72
The Nature of Gravity and How it can be Influenced	75
The Effect of Weightlessness on the Human Organism	78
The Physical Behavior of Objects when Gravity is Missing	80
Without Air	88
Perpetual Silence Prevails in Empty Space	89
Sunshine during Nighttime Darkness	89
Unlimited Visibility	90
Without Heat	91
Designing the Space Station	93
The Solar Power Plant	95

Supplying Light	96
Supplying Air and Heat	97
Water Supply	98
Long Distance Communications	98
Means of Controlling the Space Station	99
Partitioning the Space Station into Three Entities	101
The Habitat Wheel	102
The Observatory and the Machine Room	108
Providing for Long-Distance Communications and Safety	112
Partitioning the Space Station into Two Entities	112
The Space Suit	113
The Trip to the Space Station	115
Special Physical Experiments	118
Telescopes of Enormous Size	119
Observing and Researching the Earth's Surface	119
Exploring the Stars	120
A Giant Floating Mirror	121
The Most Dreadful Weapon	122
To Distant Celestial Bodies	123
The Technology of Space Travel	125
Launching from the Earth's Surface	128
The Space Station as a Base for Travel into Deep Space	128
The Attainability of the Neighboring Planets	129
Distant Worlds	131
Will It Ever be Possible to Reach Fixed Stars?	134
The Expected Course of Development of Space Travel	137
Final Remarks	140
Index	141

List of Illustrations

Figure 1, Force of Gravity..3
Figure 2, Inertial Force..4
Figure 3, Centrifugal Force..4
Figure 4, Gravitational Effects...5
Figure 5, Circular Free Orbit...6
Figure 6, Various Free Orbits...7
Figure 7, Orbiting Velocity..8
Figure 8, Earth's Atmosphere...9
Figure 9, Altitude and Air Density...10
Figure 10, Giant Cannon for Firing at the Moon..............................11
Figure 11, Reactive Force...12
Figure 12, Action and Reaction..13
Figure 13, Center of Gravity of a Rifle and Projectile......................13
Figure 14, Reaction Vehicle...14
Figure 15, Fireworks Rocket..15
Figure 16, Velocity of Expulsion..18
Figure 17, Travel Velocity..18
Figure 18, Travel and Exhaust Velocity...21
Figure 19, Vertical Ascent..25
Figure 20, Flat Ascent..26
Figure 21, Forward Thrust during Ascent..29
Figure 22, Inertial Forces during Ascent...30
Figure 23, Acceleration Polygon..31
Figure 24, Combustion Chamber and Nozzle..................................34
Figure 25, Propulsion of the Rocket..35
Figure 26, Mass and Velocity of the Rocket....................................36
Figure 27, Oberth's Alcohol and Hydrogen Rockets.......................42
Figure 28, Oberth's Booster Rocket...43
Figure 29, Launching a Rocket from Dirigibles...............................43
Figure 30, Ascent of Oberth's Small Rocket....................................44
Figure 31, Oberth's Large Rocket..45
Figure 32, Ascent of Oberth's Large Rocket....................................46
Figure 33, Oberth's Propulsion System...47
Figure 34, Hohmann's Space Rocket...48
Figure 35, Sizes of Hydrogen and Benzene Low Velocity
    Rockets..52
Figure 36, Sizes of Hydrogen and Benzene High Velocity
    Rockets..52
Figure 37, Landing with Reaction Braking......................................54
Figure 38, Landing using Air Drag Braking.....................................55

Figure 39, Hohmann's Landing Maneuver..56
Figure 40, Centrifugal Force..56
Figure 41, Excessive Travel Velocity..57
Figure 42, Operating Characteristics of Aircraft Wings.........................57
Figure 43, Wings on a Space Ship...58
Figure 44, Landing in "Forced Circular Motion"....................................59
Figure 45, Landing in "Braking Ellipses"..60
Figure 46, Inclined Trajectory...63
Figure 47, Angle of Departure...63
Figure 48, Range and Descent Velocity...64
Figure 49, Express Flight at Cosmic Velocity with Artificial Braking........66
Figure 50, Express Flight at Cosmic Velocity Avoiding Artificial
  Braking...66
Figure 51, Partly Powered Flight at Cosmic Velocity.............................67
Figure 52, Velocity and Braking for a Rocket Airplane.........................69
Figure 53, Gliding Flight without Power or Artificial Braking............70
Figure 54, Geosynchronous Orbit...72
Figure 55, Satellite as Pinnacle of Giant Tower.......................................73
Figure 56, Carousel..75
Figure 57, Giant Centrifuge..76
Figure 58, Operation of Centrifuge...77
Figure 59, Forces affecting a Free Falling Object.....................................77
Figure 60, Room in Space Station during Weightlessness.....................80
Figure 61, Writing in the weightless State..81
Figure 62, Water in a Bottle in the Absence of Gravity.........................84
Figure 63, Mercury in a Bottle in the Absence of Gravity....................84
Figure 64, Emptying a Bottle in a Weightless State...............................84
Figure 65, Emptying a Bottle through Centrifugal Force......................85
Figure 66, Behavior of Escaping Water in the Absence of Gravity.....85
Figure 67, Containers for Water in the Absence of Gravity.................86
Figure 68, Filling a Water Vessel in the Weightless State.....................86
Figure 69, Using Pressure to Empty a Water Vessel..............................87
Figure 70, Rubber Container as Water Vessel...87
Figure 71, Heating by Solar Radiation..91
Figure 72, Avoiding Heat Loss by Radiation...92
Figure 73, Concentrating the Sun's Rays by a Concave Mirror............92
Figure 74, Cooling an Object in Space via Surface Finish.....................93
Figure 75, Protecting an Object from Solar Radiation by Using
  a Mirror..93
Figure 76, Air Lock Connecting the Space Station with
  Empty Space..94
Figure 77, Solar Power Plant of Space Station...95
Figure 78, Lighting Hatch...96

Figure 79, Mirror for Reflection Solar Radiation on Window................97
Figure 80, Ventilation System for Space Station....................98
Figure 81, Thruster Motor...99
Figure 82, Orientation of Attitude Control Motors..............100
Figure 83, Velocity of Rotation and Centrifugal Force..........102
Figure 84, The Habitat Wheel....................................103
Figure 85, Directional Relationships in Habitat Wheel..........104
Figure 86, Rotating Air Lock of Habitat Wheel..................105
Figure 87, Cable Connection of Habitat Wheel...................106
Figure 88, Habitat Wheel Staircase..............................107
Figure 89, Side of Habitat Wheel Facing the Sun................108
Figure 90, Shadow Side of Habitat Wheel........................109
Figure 91, Observatory..110
Figure 92, Tube Connecting Observatory and
    Machine Room..110
Figure 93, Machine Room...111
Figure 94, Complete Space Station...............................117
Figure 95, Igniting a Piece of Wood Using a Concave Mirror.....122
Figure 96, Eight Planets of Our Solar System...................124
Figure 97, Orbits of Mercury, Venus, Earth, and Mars...........124
Figure 98, Acceleration, Deceleration, and Orbit...............127
Figure 99, Transfer Orbits......................................127
Figure 100, Uniform Acceleration and Deceleration..............136

Foreword

Frederick I. Ordway III

I have read the first full English translation of Noordung's *Das Problem der Befahrung des Weltraums* with a sense of both relief and satisfaction—relief that the book is finally accessible to the English-reading public and satisfaction that the project was achieved in the first place. Like most efforts that are ultimately realized, from my perspective the Noordung translation had quite a history. I first became aware of Noordung's work through the pages of some old pulps. In my early teens I had begun to collect science fiction magazines and books as well as non-fiction works on rocketry, the Moon and planets, and the possibility of voyages to them. In fact, I became–and continue to be–a collector.[1]

Within a few years of beginning my pulp collection, I located some early Hugo Gernsback-edited *Science Wonders Stories* that contained translations by Francis M. Currier[2] of portions from Noordung's original German-language book.[3] Particularly exciting was a full-color painting of Noordung's space station array on the cover of the August 1929 number by well-known science-fiction illustrator Frank R. Paul. That image stuck in my mind for decades.

The years passed and my only further exposure to Noordung was an occasional reference to him in my evolving non-fiction collection of books on rocketry and spaceflight.[4] Then, during the summer of 1956, while employed at Republic Aviation Corporation's Guided Missile Division, I began to ponder the feasibility of arranging for Noordung to be translated and published in the United States. Accordingly, on 23 July I wrote

---

1. Frederick I. Ordway III, "Collecting Literature in the Space and Rocket Fields," *Space Education* supplement to the *Journal of the British Interplanetary Society* 1 (September 1982): 176-82; 1 (October 1983): 279-87; and 1 (May 1984): 326-30. Also "The Ordway Aerospace Collection at the Alabama Space and Rocket Center," in *Special Collections: Aeronautics and Space Flight Collections*, edited by Catherine D. Scott (New York: Haworth Press, 1985), pp. 153-72.
2. Hermann Noordung, "The Problems of Space Flying," translated by Francis M. Currier *Science Wonder Stories* 1 (July 1929): 170-80; (August 1929): 264-72; and (September 1929): 361-368.
3. Hermann Noordung, *Das Problem der Befahrung des Weltraums: Der Raketen-Motor* (Berlin: Richard Carl Schmidt & Co, 1929).
4. I learned about Noordung's space station concept not only from the old *Science Wonder Stories* but from reading the first edition of Willy Ley's *Rockets: The Future of Travel Beyond the Stratosphere* (New York: Viking, 1944), pp. 225 and 229. By the time Ley's book had been expanded into *Rockets, Missiles and Space Travel* (New York: Viking, 1951), Noordung had been relegated to a footnote on page 332 and a bibliographic citation on page 420. Years later, with the appearance of *Rockets, Missiles, and Men in Space* (New York: Viking, 1969), a couple of pages–300 and 301–were devoted to Noordung. And that's as far as

to the German publisher Richard Carl Schmidt & Co. in Berlin. Within a couple of weeks I received a reply from the company's head office in Braunschweig to the effect that "We are in pri[n]ciple ready to assist your project." The letter went on to say that "Before having a final decision please give us the ad[d]ress of Mr. Hugo Gernsba[c]k, with whom we should like to have priorly a contact".[5]

As it happened, I had anticipated this requirement and was already in touch with Gernsback at his *Radio-Electronics* magazine office located at 154 West 14th Street in New York City. I had hoped that a full translation of Noordung might have been made by Currier even though Gernsback had only published portions. After some correspondance and telephone conversations, he advised me on 5 July 1956 that the original English translation "...is no longer available, as we have no records going that far back at the present time." He went on to explain that "...I have talked to several people about this, but so far have not been able to get any further information on it." We later got together for lunch at which time he reconfirmed that a complete English translation did not exist; in fact, he had no surviving files on the matter. This meant that we would have either to make do with the extracts published in the old *Science Wonder Stories* or have the book translated anew.

The month following my conversations with Hugo Gernsback, I made what turned out to be a career-changing trip to Redstone Arsenal in Huntsville, Alabama, for meetings with Wernher von Braun and members of his team. Then, in December von Braun spent a couple of days with my wife and me at our home in Syosset, Long Island. It was then that arrangements were made for me to join him at Redstone. So, in February 1957, with wife, two children and collie dog appropriately name Rocket, we pulled up stakes for the move to Huntsville. Noordung was no longer a concern.

Willy Ley went. Based to a large extent on Ley's limited notations, similarly brief mention of Noordung appeared in other books over the years, including some coauthored by me, e.g. Frederick I. Ordway III and Ronald C. Wakeford, *International Missile and Spacecraft Guide* (New York: McGraw-Hill, 1960), p. 212, and Wernher von Braun and Frederick I. Ordway III, *History of Rocketry & Space Travel* (New York: Crowell, 1966), p. 202 and later editions (Crowell, 1969; Crowell, 1975; and Harper & Row, 1985—the last with the title *Space Travel: A History*). Other authors gave Noordung the same rather spares treatment, e.g. David Baker, *The History of Manned Space Flight* (New York: Crown, 1981), p. 14. In 1987, Sylvia Doughty Fries and I offered a bit more—but hardly adequate—space to Noordung in our "The Space Station: From Concept to Evolving Reality," *Interdisciplinary Science Reviews* 12 (June 1987): 143-59.

5. All correspondance referred to in this foreword is located in the Ordway Collection, Center Library and Archives, U.S. Space & Rocket Center, Huntsville, Alabama.

After our arrival at Redstone, life became so busy and so exciting that the Noordung project further faded from mind. In late May, we celebrated at Huntsville the first successful firing of a 1,500-mile-range Jupiter from Cape Canaveral; in August a 600-mile-altitude, 1,300-mile-range flight of a Jupiter-C three-stage rocket; and in January 1958 the orbiting of America's first satellite, Explorer 1. Then, in July 1960, the von Braun team transferred to NASA as the George C. Marshall Space Flight Center, in May 1961 Alan Shepard became the first American astronaut to fly into space, and later that same month President Kennedy announced a national goal of sending astronauts to the Moon within the decade.

By the time of the Apollo announcement, five years had passed since I had begun thinking about having Noordung translated into English. A couple more years went by until, during the spring of 1963, Harry O. Ruppe–who was deputy director of NASA-Marshall's Future Projects Office–and I began to bring the project back to life.

Our first step was to write to Willy Ley in Jackson Heights, New York, an old friend and colleague from my Long Island days. In a letter of 22 April, I told him that Ruppe had begun planning the translation and that we had discussed the introduction (which we both hoped Ley would write). I added that "One thing that puzzles him [Ruppe] is the almost complete absence of information on Noordung. Apparently he made every effort to conceal his identity. He had no history of publications prior to this book and made no effort to publish further after it appeared." I then posed seven questions:

1. Who was Noordung?
2. Why did he try to conceal his identity?
3. Is he known to have written anything other than this book?
4. Where did he get his training?
5. What happened to him after he wrote the book?
6. What contacts did he have with other rocket and astronautical pioneers?
7. How well did his book do and what influence did it have on the development of astronautical thinking in prewar Germany?

Agreeing that "Yes, Noordung is a difficult case," Willy Ley answered the seven questions in a letter to me dated 3 May 1963:

1. His real name was Potočnic, first name unknown. He was a captain in the Austro-Hungarian army of World War I, and since he used the title Ingenieur it is generally assumed he was a Captain in the Engineer Corps.

2. Two possible reasons (A) he might have felt that his real name, being Czech [see below], was a slight handicap. (B) he might have wished to indicate that he was speaking as an individual, not as an

officer (albeit retired) of the armed forces, just as I have used the pen name Robert Willey for the few science fiction stories I wrote in the past, to show that this was meant to be fiction. (I did not hide behind the penname, though, everybody knew it was me.) There is a third possibility, somewhat unkind to the gentleman. He might have wished to keep his name from the disbursing office for pensions so that he would receive his full pension in spite of outside income.

3. Nothing else by him is known.
4. See answer to # 1.
5. He seems to have been fairly old when he wrote the book, presumably he died a few years later.
6. Virtually none.
7. His book never got beyond its first printing, it was strongly critized by [Austrian space pioneer Guido] von Pirquet because of its errors in the tables (the one about rocket efficiency). It was one of those books where the prophetic content was not realized until much later. At the time it was new, everybody only looked at the mistakes he made.

Although Ley said in answering question No. 1 that he didn't know Potočnic's first name, a couple of years later he provided me a translation from an article on Noordung by Prof. Erich Dolezal in the Viennese magazine *Universum* (Vol. 33, Heft 2, 1965) to the effect that "The book [Noordung's] was one of the best of the early period, it deal[t] especially with the problem of the space station...supposed to be in a synchronous orbit...The pen name Hermann Noordung covered the Austrian (Army) Officer Hermann Potočnik who had been born in Pola [now, Pula] in 1892. His father had been a Navy staff surgeon who had been a participant in the naval battle of Lissa (1866)." Learning that Noordung had died on 27 August 1929 of a pulmonary disorder, Ley commented "...no wonder he [n]ever answered anything, he died during the same year his book was published."

In May 1963, Harry Ruppe began planning the translation, our artist/designer colleague Harry H-K Lange considered either reproducing or redoing the illustrations, and I got back in touch with the Richard Carl Schmidt publishers in Germany to make final arrangements to secure for us the publication rights. Replying on 24 June 1963, the company proposed that a formal agreement be drawn up and a royalty rate established. With this reasonable response in mind, we approached the McQuiddy Publishers of Nashville and Aero Publishers, Inc. of Los Angeles regarding publishing and distributing the book. A year went by, and we got nowhere; and, ultimately, the German publisher simply refused to respond to further inquiries. Since Mcquiddy and Aero understandably would not proceed without Richard Carl Schmidt's approval, we dropped all plans to publish a translation.

All the while, Ruppe's, Lange's and my workloads at the NASA-Marshall Center were increasing so we relegated the Noordung project to our inactive files. Ruppe's final notes to me echo in part Willy Ley's comments. "This guy Noordung is quite a mysterious figure. You know, usually you write a book, amongst many other things, to reap some fame. Certainly this was not one of his motives. Indeed this guy is nearly impossible to grasp. He wrote a book, he published it and then he vanished again. He is not well known prior to the book. There is nothing we know of him thereafter. He did not republish."

In a way, [Ruppe continued] he does not belong to the old pioneers and in another way he does. He is aware of the state-of-the-art of his time. His mathematical knowledge is not too strong. He misinterprets some concepts like his efficiency considerations which were quite well discussed in [Hermann] Oberth's book a couple of years prior to his own. But on the other hand, he shows very, very capable concepts. His space station design is kind of a model–well, we honestly haven't progressed very much further except for one of his quite funny mistakes, placing it in a 24 hour orbit.

And there the matter rested until 1992 when I learned that Ron Miller had obtained a copy of Noordung's book in the original Slovenian![6] Noordung, it seems, was not Czech as Ley had surmised but rather the Slovenian Herman Potočnic born on 22 December 1892 in what is now Croatia; his mother–though also Slovenian–had Czech ancestors. I quickly persuaded Miller to provide a short write-up of the appearance of Noordung/Potočnic in Slovenian for a special issue of the *Journal of the British Interplanetary Society* being edited by Frank Winter and me. This he graciously consented to do.[7]

Some months later, Noordung/Potočnic again came to my attention with the arrival from Sri Lanka of a memorandum from Arthur C. Clarke dated 15 January 1993:

PROBLEM VOŽNJE PO VESOLJU

This afternoon just as I was leaving for the Otters Club to beat up the locals at table tennis, I noticed two young European backpackers hovering around my gate. Stopped to find who they were, and discovered they were a couple of Slovenes, who'd hiked here to deliver this book

---

6. Herman Potočnic, *Problem Vožnje po Vesolju* (Ljubljana: Slovenska Matica, 1986).
7. Ron Miller, "Herman Potocnik - *alias* Hermann Noordung," *Journal of the British Interplanetary Society* 45 (July 1992), "Pioneering Rocketry and Spaceflight" issue, Part I:295-6.

to me!! Do you know it? I've never seen the original, and the illustrations are fascinating. Though of course, I was familiar with some of them, notably the space station design.

I immediately sent Clarke a copy of the Miller article; and, on 17 February 1993, he acknowledged its receipt with thanks and had this to add:

Here's an incredible coincidence–a week after the Slovenes gave me their edition of Potočnik, Luis Marden of National Geographic mailed me a copy of the German edition, with dust jacket, he'd found when cleaning out his library! I'd never seen it before, and was delighted to have it. What an incredible man Potočnik must have been, perhaps in some ways quite as remarkable as Oberth.

About the time Noordung/Potočnik was re-entering my life after so many years and false starts towards translation into English, I got to talking with NASA's Chief Historian Dr. Roger Launius. Once again a coincidence: he, himself, had been pondering the feasibility of NASA's sponsoring the translation and subsequent publication. I quickly loaned him my files hoping they might be of some use. In due course, under the guidance of Dr. John Dillard Hunley of the History Office, a professional translation was finally made, Dr. Ernst Stuhlinger of Huntsville agreed to critically review it, and publication subsequently was realized. Mission accomplished!

Preface

J.D. Hunley

Hermann Noordung's *Das Problem der Befahrung des Weltraums*, published here in English translation, was one of the classic writings about spaceflight. Its author, whose real name was Herman Potočnik, was an obscure former captain in the Austrian army who became an engineer. He was born on 22 December 1892 in Pola (later, Pula), the chief Austro-Hungarian naval station, located on the Adriatic in what is today Croatia. As the location might suggest in part, his father served in the navy as a staff medical officer. The name Potočnik is Slovenian, also the nationality of Herman's mother, who had some Czech ancestors as well. The young man was educated in various places in the Habsburg monarchy, attending an elementary school in Marburg (later, Maribor) in what is today Slovenia. He enrolled in military schools with emphases on science and mathematics as well as languages for his intermediate and secondary schooling, in the obscure town of Fischau, Lower Austria, and in Mährisch-Weißkirchen (later the Czech city of Hranice), respectively.

Hermann Noordung, alias for Herman Potočnik

Following that, he attended the technical military academy in Mödling southwest of Vienna. Upon graduation, he received his commission as a lieutenant in the Austro-Hungarian army, where he served during World War I in a railroad (guard) regiment. From 1918 to 1922 he studied electrical engineering at the Technical Institute in Vienna, although tuberculosis had forced him to leave the army in 1919. While he appears to have set up a practice as an engineer, his illness evidently prevented him from working in that capacity. But he did become interested in the spaceflight movement. He contributed monetarily to the journal of the German Society for Space Travel (Verein für Raumschiffahrt or VfR), *Die Rakete* (The Rocket), begun in 1927, and he corresponded with Hermann Oberth (1894-1989), whose book *Die Rakete zu den Planetenräumen* (The Rocket into Interplanetary Space, published in 1923) essentially launched the spaceflight movement in Germany and laid the theoretical foundations for future space efforts there. Another correspondent was Baron Guido von Pirquet (1880-1966), who wrote a series of articles on interplanetary travel routes in *Die Rakete* during 1928 that suggested space stations as depots for supplying fuel and other necessaries to interplanetary rockets. The rockets, in his conception, would be launched from the stations rather than from the Earth to avoid the amounts of propellants required for escape from the home planet's gravitational field, which would be much weaker at the distance of a few hundred kilometers.

Oberth encouraged Potočnik to express his ideas about rocketry and space travel in a book, which he completed with its 100 illustrations in 1928. Potočnik's gratitude to Oberth and the enthusiasts around him in Germany led the still young but ailing engineer to assume the pen name of Noordung (referring to the German word for north, *Nord*) in honor of the fellow space enthusiasts to his north. He published the book with Richard Carl Schmidt & Co. in Berlin in 1929, only to die soon afterwards on 27 August 1929 of tuberculosis.[1]

Potočnik's book dealt, as its title suggests, with a broad range of topics relating to space travel, although the rocket motor that forms the book's subtitle was not especially prominent among them. What makes the book important in the early literature about space travel is its extensive treatment of the engineering aspects of a space station. Potočnik was hardly the first person to write about this subject, as the comment about Pirquet above would suggest. The idea in fictional form dates back to 1869-1870 when American minister and writer Edward Everett Hale (1822-1909) published "The Brick Moon" serially in *The Atlantic Monthly*. The German mathematics teacher, philosopher, and historian of science Kurd Laßwitz (1848-1910) followed this up in 1897 with his novel *Auf zwei Planeten* (translated into English as "Two Planets" in 1971), which featured a Martian space station supported by antigravity that served as a

---

1. Frank H. Winter, "Observatories in Space, 1920s Style," *Griffith Observer* 46 (Jun. 1982): 3-5; Fritz Sykora, "Pioniere der Raketentechnik aus Österreich," *Blätter für Technikgeschichte* 22 (1960): 189-192, 196-199; Ron Miller, "Herman Potocnik—*alias* Hermann Noordung," *Journal of the British Interplanetary Society* 45 (1992): 295-296; Harry O. Ruppe, "Noordung: Der Mann und sein Werk," *Astronautik* 13 (1976): 81-83; Herbert J. Pichler, "Hermann Potočnik-Noordung, 22. Dez. 1892-27 Aug. 1929," typescript from a folder on Potočnik in the National Air and Space Museum's Archives. These sources on Potočnik's life agree in the essentials but disagree in some particulars, even to the spelling of his first name, which appears as Herman (with one n) on the title page of the Slovenian edition of his book, first published in his native language in 1986. Winter and Sykora also discuss both Oberth and Pirquet. Further information about both appears in Barton C. Hacker, "The Idea of Rendezvous: From Space Station to Orbital Operations in Space-Travel Thought," *Technology and Culture* 15 (1974): 380-384. Other sources on Oberth's life include Hans Barth, *Hermann Oberth: "Vater der Raumfahrt"* (Munich: Bechtle, 1991) and John Elder, "The Experience of Hermann Oberth," as yet unpublished paper given at the 42nd Congress of the International Astronautics Federation, October 5-11, 1991, Montreal, Canada. On Pirquet's series of articles, see also *Die Rakete: Offizielles Organ des Vereins für Raumschiffahrt E.V. in Deutschland* 2 (1928): esp. 118, 137-140, 184, 189.

2. Cf. Winter, "Observatories in Space," pp. 2-3; Hacker, "Idea of Rendezvous," pp. 374-375.

staging point for space travel.² Two years before the appearance of Laßwitz's book, the earliest of the recognized pioneers of spaceflight theory, the Russian school teacher Konstantin E. Tsiolkovsky (1857-1935), published a work of science fiction entitled (in English translation) *Reflections on Earth and Heaven and the Effects of Universal Gravitation* (1895) in which he discussed asteroids and artificial satellites as bases for rocket launches. He also discussed the creation of artificial gravity on the manmade space stations through rotation.³ Unlike Laßwitz, Tsiolkovsky was not content with science fiction, however. Between 1911 and 1926, the Russian spaceflight theorist expanded his ideas and subjected them to mathematical calculation. In the process, he elaborated a concept of a space station as a base for voyages into space but did not develop it in any detail.⁴ Others in what was then the Soviet Union also developed ideas about space stations,⁵ but they were little known in the West. Thus, for the development there of conceptions about space stations the writings of Oberth were much more important. The Romanian-German spaceflight theorist wrote briefly about "observation stations" in his 1923 book and discussed some of their possible uses such as observation and military reconnaissance of the Earth, service as a fueling station, and the like.⁶ In the expanded and more popular version of his book published in 1929, *Wege zur Raumschiffahrt* (translated as *Ways to Spaceflight*), Oberth covered these ideas in more detail, but he devoted most of his attention to a space mirror that could reflect solar energy upon a single point on Earth or upon a wider region for keeping northern ports free of ice in winter, illuminating large cities at night, and other applications.⁷

3. Hacker, "Idea of Rendezvous," pp. 375-376; N.A. Rynin, *Interplanetary Flight and Communication*, III, 7, *K.E. Tsiolkovskii, Life, Writings, and Rockets* (Leningrad, 1931), trans. by Israel Program for Scientific Translations (Jerusalem: NASA TT F-646, 1971), pp. 24-25; I.A. Kol'chenko and I.V. Strazheva, "The Ideas of K.E. Tsiolkovsky on Orbital Space Stations," *Essays on the History of Rocketry and Astronautics: Proceedings of the Third Through the Sixth History Symposia of the International Academy of Astronautics*, ed. R. Cargill Hall (Washington, D.C.: NASA Conference Publication 2014, 1977), vol. 1, p. 171. This last work has since been reprinted as *History of Rocketry and Astronautics*, AAS History Series, vol. 7, pt. 1 (San Diego: Univelt, 1986).
4. Hacker, "Idea of Rendezvous," pp. 376-377; K. E. Tsiolkovskiy (*sic*), "The Exploration of the Universe with Reaction Machines," in *Collected Works of K.E. Tsiolkovskiy*, vol. 2, *Reactive Flying Machines*, ed. B. N. Vorob'yev et al., trans. Faraday Translations (Washington, D.C.: NASA TT F-237, 1965), pp. 118-167, esp. 146-154; Tsiolkovskiy, "Exploration of the Universe with Reaction Machines," in *Reactive Flying Machines*, pp. 212-349, esp. pp. 338-346; Kol'chenko and Strazheva, "Ideas of Tsiolkovsky on Orbital Space Stations," pp. 172-174.
5. Hacker, "Idea of Rendezvous," pp. 377-379.
6. Hermann Oberth, *Die Raketen zu den Planetenräumen* (Nürnberg: Uni-Verlag, 1960; reprint of 1923 work), pp. 86-89; for a more detailed description of Oberth's ideas, see Winter, "Observatories in Space," pp. 3-4.
7. Hermann Oberth, *Ways to Spaceflight*, trans. Agence Tunisienne de Public-Relations (Washington, D.C.: NASA TT F-622, 1972), pp. 477-506.

Oberth's expanded book appeared, according to Frank H. Winter, immediately after that of Potočnik.[8] Winter does not reveal his source for this information, but its accuracy appears to be validated by the numerous references to Noordung in *Ways to Spaceflight*.[9] On the other hand, since a review of Potočnik's book appeared in the October 1928 issue of *Die Rakete*, it is possible that Oberth saw an advance copy.[10] While these other early works on space stations had important theoretical influences, what Potočnik's book offered that they didn't was engineering details about how a space station might be constructed.[11]

While Potočnik's book is clearly the classic statement of how a space station might be constructed,[12] it is difficult to know how to assess its real importance for the later design, construction, and use of space stations and other spacecraft. For one thing, the work received considerable criticism, apparently even before it was published.[13] The unsigned October 1928 review in *Die Rakete* praised the book as a successful and understandable introduction to the highly interesting problem of spaceflight. But it said the work paid too little attention to recent contributions to the topic, such as those that had appeared in *Die Rakete* itself. The author's treatment of the issue of [rocket] efficiency could be accepted only with caution, the review went on. Noordung presented a detailed treatment of a space station, but he placed it in a 35,000 kilometer, geostationary orbit, which was not practical according to the existing state of research (see below).[14]

As can be seen from the translation that follows (p. 74), in fact Potočnik spoke of a 35,900 kilometer orbit but also discussed the possibilities of orbits at different distances from the Earth and at other inclinations than the plane of the equator. Thus, this particular criticism was a bit unfair.

Both Willy Ley and Pirquet also criticized the book shortly after it appeared.[15] Their arguments appear to have been conveniently summarized by Ley in his widely-circulated *Rockets, Missiles, and Space Travel*,

8. "Observatories in Space," p. 4.
9. Pp. 56, 138, 210, 247, 302, 308, 309, 351, 352, 354, 417, 478.
10. *Die Rakete* 2 (Oct. 1928): 158-159.
11. Sykora, "Pioniere der Raketentechnik," p. 198; Frank Winter, *Rockets into Space* (Cambridge, Mass.: Harvard University Press, 1990), pp. 25-26. In fact, Winter states, "his work was so comprehensive that no other space station study appeared until the mid-1940s."
12. Cf. Winter, "Observatories in Space," p. 4.
13. Willy Ley (*Rockets, Missiles, and Men in Space* [New York: Viking, 1968], p. 540) and others have dated Potočnik's book from 1928, but as Harry Ruppe points out ("Noordung," p. 82n) the 1929 edition of the book gives no indication that it is a second printing. Moreover, the copyright dates from 1929.
14. *Die Rakete* 2 (Oct. 1928): 158-159.
15. Ruppe, "Noordung," p. 82.

which had appeared under a variety of titles and revisions and some twenty printings from 1944 to 1968.[16] In the 1961 revised and enlarged edition of that book Ley recalled that Potočnik had "succeeded in getting himself into the bad graces of the few rocket men of the time by a number of peculiarities. The first of them...was a somewhat fantastic method of calculating the over-all efficiency of a rocket." This particularly "incensed" Pirquet, Ley said. Secondly, he failed to answer correspondence—naturally enough since he was dying of tuberculosis—but most of his contemporaries were evidently ignorant of that fact. The third peculiarity was his insistence that the space station must be in a 24-hour orbit, something that would decrease its value by, say, 75 per cent. In such an orbit it could observe only one hemisphere of the earth and that one not very well because of the great distance, which also would make the station's construction and maintenance rather expensive in terms of extra tons of fuel consumed [on the trip to the station]. He did have a number of interesting ideas, but each one of them came out somewhat flawed. Nevertheless, Ley remembered writing to Pirquet "that 'Noordung's plans are of great historical significance even now.'"[17]

Ley went on to note several "essentially correct" thoughts in Potočnik's design, including an airlock and the plan to obtain power from the sun. But he said there were also "strange mistakes," such as an excessive concern with heating the station when what it really needed was air conditioning because of the absorbed heat from the sun, the heat from the bodies of the crew, and that generated by electric motors. Also, Potočnik wanted to rotate his habitat wheel—the occupied element among the three connected but separate units in his space station—every eight seconds so as to create a full g of artificial gravity. Ley said that $1/3$ g would be adequate and would allow the station to be "lighter and therefore cheaper to carry into an orbit piecemeal."[18]

Besides such criticisms, which must have served to reduce the influence of Potočnik's detailed designs, there was also the problem that the book appears not to have been widely available to the non-German speaking world. There was a very early, partial English translation by Francis M. Currier that appeared serially in what might seem (although it really is not) a strange place, Hugo Gernsback's *Science Wonder Stories* in mid-

---

16. As stated on the back of the title page of the 1968 edition, *Rockets, Missiles, and Men of Space*, which had a less extensive discussion of the author and book than the 1961 edition referred to in the narrative above and cited in the next footnote. Interestingly, the 1951 edition, also entitled *Rockets, Missiles, and Space Travel* (like the 1961 edition), relegated Potočnik and the discussion of his book to a footnote, where none of the criticism appeared.
17. *Rockets, Missiles, and Space Travel* (New York: Viking, 1961), p. 369.
18. *Ibid.*, pp. 369-370, 375.

1929,[19] but it is uncertain how many people might have read the work in that magazine. Moreover, even if they did read it, how many of them kept their copies and how available the partial translation was in most libraries are questions no one seems to have asked.[20] There was also a (partial?) translation done for the British Interplanetary Society and kept at its library in London, apparently as an unpublished typescript.[21] And a Russian translation appeared about 1935 but may have been only partial since it was only 92 pages long compared with the 188 in the original German edition.[22] Despite these translations, at least in Britain the work

19. *Science Wonder Stories* 1 (1929): 170-80, 264-72, and 361-68 (July-September issues). Photocopies in the National Air and Space Museum's archives in a folder on Potočnik; copies of the original magazine at the Library of Congress. *Science Wonder Stories* is not as strange a place for a translation of an engineering study to appear as the title might suggest. Gernsback (1884-1967) carried on the masthead of the new magazine the dictum, "Prophetic Fiction is the Mother of Scientific Fact," and it was his policy to publish only stories that were scientifically correct, with a "basis in scientific laws as we know them. ..." He also carried a section entitled "Science News of the Month," in which the publication sought to report up-to-date scientific achievements "in plain English." Copy of editorial page from *Science Wonder Stories* 1 (1929) in Hugo Gernsback, biographical file, NASA Historical Reference Collection. See also Tom D. Crouch, "'To Fly to the World in the Moon': Cosmic Voyaging in Fact and Fiction from Lucian to Sputnik," in *Science Fiction and Space Futures, Past and Present*, ed. Eugene M. Emme, AAS History Series, Vol. 5 (San Diego: Univelt, 1982), pp. 19-22, and Sam Moscowitz, "The Growth of Science Fiction from 1900 to the Early 1950s," in *Blueprints for Space: Science Fiction to Science Fact*, ed. Frederick I. Ordway III and Randy Liebermann (Washington, D.C.: Smithsonian Institution Press, 1992), pp. 69-82, among other sources on Gernsback, who is generally credited with coining the term "science fiction."
20. As of 17 March 1994, the On-Line Computer Library Center (OCLC) data base showed 8 libraries in the U.S. as holding hard copies of the June 1929-May 1930 issues of *Science Wonder Stories* plus 14 others with those issues on microfilm. There may, of course, have been other libraries holding the set including the partial Potočnik translation at earlier dates, but many of the current holdings of microfilm especially seem to have been acquired recently. The *Library of Congress Pre-1956 Imprints* shows only 6 libraries holding *Science Wonder Stories* and its sequels, for example, and Donald H. Tuck (*The Encyclopedia of Science Fiction and Fantasy*, 3 vols. [Chicago: Advent Publishers, Inc., 1974-1982], vol. 1, p. 185) reports that all of Gernsback's science fiction magazines "led checkered careers, some lasting only brief periods." On the other hand, Fred Ordway reports that he owns a fine set of *Science Wonder Stories* and that many copies of such pulp magazines found their way to England as ballast on ships returning less than full from carrying certain types of cargo to the United States. Thus, there may be many copies of the publication still in private hands.
21. Hacker, "Idea of Rendezvous," p. 384n. Hacker seems to suggest that this was a full translation, but according to L. J. Carter, long-time executive secretary of the BIS, this was not the case. In a letter to Fred Ordway on 15 April 1994, he wrote in answer to his question about the existence of such a document, "yes, we used a translation of extracts from it [the Potočnik book] when considering the early BIS Space Station designs. This, however, was done before the war so it is unlikely that anything has survived." Information kindly provided by Fred Ordway.
22. On-Line Computer Library Center printout from early March 1994 showing one copy in the United States at the California Institute of Technology.

appears not to have been readily available. For example, the famous British science fiction writer and member of the British Interplanetary Society Arthur C. Clarke had cited Potočnik's book in his October 1945 article in *Wireless World* where he had discussed using a satellite in geosynchronous orbit for purposes of radio communications.[23] But he later stated that at the time he had not seen the book, only pictures of Potočnik's space station in science fiction magazines. He only obtained his "first copy of Potočnik's classic book" in 1993. And while some Austrians have tried to credit Potočnik with first conceiving of both the communications satellite and a geostationary orbit for it, Clarke pointed out that Tsiolkovsky had written about the geostationary orbit at the beginning of the century and that Oberth first wrote about using space stations for communications in his 1923 book, although through optical means rather than radio. Clarke nevertheless credited Potočnik with envisioning the use of short waves for communications between the Earth and his space station.[24]

Evidently, Clarke was not aware of the translation in the British Interplanetary Society's library. Other members of the society apparently were not either, because a brief notice appeared in its publication, *Spaceflight*, in 1985 announcing that the editors had "at last" secured a copy of the Potočnik book in the original German. "This was a book to which the pre-war BIS Technical Committee frequently referred," the notice went on, "though none had actually seen it, for the simple reason that no copies were available."[25]

Both Frank Winter and Adam Gruen have suggested that Potočnik's book formed the basis for a plotless short story entitled "Lunetta" that Wernher von Braun wrote in 1929, describing a trip to a space station.[26]

---

23. Entitled "Extra-Terrestrial Relays," this article is generally credited with being the first account that clearly outlined the idea of communications satellites. In it, Clarke suggested three satellites spaced equidistantly at altitudes of 22,300 miles (35,900 kilometers) to provide complete coverage of the Earth. See, e.g., *Telecommunications Satellites: Theory, Practice, Ground Stations, Satellites, Economics*, ed. K[enneth] W. Gatland (London: Prentice-Hall, Inc., 1964), pp. 1-2, 21; Neil McAleer, *Odyssey: The Authorized Biography of Arthur C. Clarke* (London: Gollancz, 1992), pp. 58-61.
24. Clarke's comments appeared in a letter to the *IEEE Spectrum* 31, (Mar. 1994): 4, responding to a letter from the Austrian, Viktor Kudielka, in the same journal, vol. 30, (June 1993): 8, where the latter claimed precedence for Potočnik in inventing a geostationary orbit for short wave communications. (My thanks to Lee Saegesser of the NASA History Office for bringing the Clarke letter to my attention.) Pichler, "Potočnik-Noordung," pp. 2, 10, also claims precedence for Potočnik in inventing communications satellites and the geosynchronous orbit.
25. "Space Station Update," *Spaceflight* 27 (Feb. 1985): 92.
26. Winter, *Prelude to the Space Age: The Rocket Societies* (Washington, D.C.: Smithsonian Institution Press, 1983), p. 114; Gruen, "The Port Unknown: A History of the Space Station *Freedom* Project" (as yet unpublished typescript dated 30 April 1993 and submitted to the NASA History Office), p. 13n7.

If correct, this hypothesis would suggest an important link in the evolution about ideas for a space station. As is well known, von Braun—technical director of the German rocket development center at Peenemünde that developed the V-2 ballistic missile during World War II and later director of NASA's Marshall Space Flight Center while it developed the Saturn V rocket—wrote an article for the popular *Collier's* magazine in 1952 in which he described a space station at least superficially similar to Potočnik's.[27] This article, others in the eight-part *Collier's* series of which it was a part, and a Walt Disney television series, *Man in Space*, in which von Braun, Ley, and others appeared, helped establish the American popular image of a space station and of space exploration as well as a vision of a space station that, in Howard McCurdy's words, "would continue to guide NASA strategy through the decades ahead."[28] As McCurdy also stated, "More than any other person, von Braun would be responsible for clarifying in the American mind the relationship between space stations and space exploration."[29]

Thus, if Potočnik indeed influenced von Braun, through the latter he must also have influenced the United States and NASA. Unfortunately, there appears to be no conclusive evidence for such influence on Potočnik's part. Winter quotes von Braun as saying that during the period around 1929 he "read everything in the space field, including Willy Ley's popularizations,"[30] and it seems likely that he would have read the Potočnik book along with the other contemporary literature. As stated above, von Braun wrote "Lunetta" in 1929, so it could easily have been influenced by Potočnik. At the time the future rocket engineer and manager was still attending secondary school at the Hermann Lietz International School on the island of Spiekeroog in the North Sea. The school published the plotless story in its publication, *Leben und Arbeit* (life and

27. "Crossing the Last Frontier," *Collier's* Mar. 22, 1952: 24-29, 72-73.
28. Howard E. McCurdy, "The Possibility of Space Flight: From Fantasy to Prophecy," paper delivered at the annual meeting of the Society for the History of Technology, 15 Oct. 1993, pp. 2, 10-11, 16; *The Space Station Decision: Incremental Politics and Technological Choice* (Baltimore: Johns Hopkins University Press, 1990), pp. 5-8, quotation from p. 8. Cf. the comment of W. Ray Hook, who had helped develop space station concepts at Langley Research Center and was later the manager of its Space Station Office and then its Director for Space. In an apparently unpublished paper entitled "Historical Review and Current Plans," p. 1, received in late 1983 in the NASA History Office and now residing in a folder marked "Space Station Historical" in the NASA Historical Reference Collection, he stated of von Braun and others' space station concept published in *Collier's* in 1952, "The basic tenets and objectives of this proposal were essentially sound and have been pursued with varying levels of activity ever since." See also in this connection, Randy Liebermann, "The *Collier's* and Disney Series," *Blueprint for Space*, pp. 135-146, and the video on the subject at the Smithsonian National Air and Space Museum.
29. *Space Station Decision*, p. 5.
30. Winter, *The Rocket Societies*, p. 114.

work) in the 1930-1931 issue, volume 2-3 on pp. 88-92.[31] It describes a trip by rocket to a space station and back, with some of the details similar to those in Potočnik's book, which does contain a description of a somewhat similar trip (pp. 115-118 of the translation).[32] However, Oberth had also included a description of a rocket flight through interplanetary space in *Ways to Spaceflight*, upon which von Braun could have loosely based his own story.[33] Thus, the influence of Potočnik's book upon von Braun remains probable but speculative.

What can be stated unequivocally is that Potočnik's book was widely known even to people who may have seen only photographs or sections from the book in translation. For example, Harry E. Ross of the British Interplanetary Society proposed a large, rotating space station in conjunction with Ralph A. Smith in 1948-1949, basing it upon Potočnik's drawings although neither of them could read his German.[34] It is also clear that although the details of Potočnik's designs for a space station were not repeated in later space stations, he foresaw many of the purposes for which space stations as well as other spacecraft were used. As the reader of the translation will discover, Potočnik, following Oberth on many points, predicted a great many uses for his space station. These included physical and chemical experiments conducted in the absence of gravity and heat, studies of cosmic rays, astronomical studies without the interference produced by the Earth's atmosphere, studies of the planet Earth itself from the vantage point of space (including meteorological and military applications of the resulting information), the use of a space mirror to focus the rays of the sun upon the Earth for a variety of purposes (including combat), and use as a base for travelling further into space (pp. 118-130).

Not all of these goals have been implemented. But on Skylab, which provided an orbiting habitat for three 3-person crews between May 1973

---

31. Ernst Stuhlinger and Frederick I. Ordway, *Wernher von Braun: Aufbruch in den Weltraum* (Munich: Bechtle Verlag, 1992), pp. 47-48. The English version of this book, *Wernher von Braun: Crusader for Space*, vol. 1: *A Biographical Memoir* (Melbourne, Florida: Krieger Publishing Company, 1994), contains this information on p. 16.
32. Copy of "Lunetta" from *Leben und Arbeit* 2/3 (1930/31): 88-92 in von Braun biographical folder, "Sputnik to Dec. 1965," in the NASA Historical Reference Collection.
33. *Ways to Spaceflight*, pp. 410-435.
34. Hacker, "Idea of Rendezvous," pp. 384-385; Frederick I. Ordway, III, "The History, Evolution, and Benefits of the Space Station Concept (in the United States and Western Europe)," paper presented at the XIIIe Congress of the History of Science, Section 12, Moscow, August 1971, p. 6, later published in the *Actes du XIIIe Congrès International d'Histoire des Sciences* 12 (1974): 92-132. As Hacker points out, it was for Ross and Smith that the B.I.S. English translation was later prepared. The design of their space station appeared in the London *Daily Express* in November 1948 and later as "Orbital Bases" in the *Journal of the British Interplanetary Society* 8 (Jan. 1949): 1-19.

and February 1974, experiments included various solar studies, stellar astronomy, space physics, experiments to study the Earth, materials science, zero-gravity studies, and studies of radiation, among others.[35] Numerous other spacecraft from the Hubble Space Telescope to Shuttle orbiters equipped with a spacelab have also fulfilled some of the expectations set forth in a general way by Potočnik.[36]

It is also quite possible that by proposing the first actual design for a space station and by offering illustrations of that design, Potočnik helped to fixate the imagination of people interested in spaceflight upon a space station as an important goal in itself and a means to the end of interplanetary flight. Since 1959 NASA has conducted at least a hundred studies of space station designs,[37] and the idea of a space station became a firm fixture in NASA's planning from the mid-1960s to the present day.[38] Much more continuously than the United States, the former Soviet Union and Russia have had a space station program dating back to the launch of Salyut 1 on 19 April 1971.[39] In view of the mid-1930s translation of Potočnik's book into Russian, perhaps that program, too, owes something to the little-known Austro-Hungarian engineer's study.

In view of these possibilities, it is time for Potočnik's book to be readily available to the English-speaking world in a readable and accurate translation. At the suggestion of Lee Saegesser, the NASA History Office commissioned the translation that appeared as NASA TT-10002 in 1993. The NASA STI Program had that translation done by SCITRAN of Santa Barbara, California, and subsequently edited by Jennifer Garland of the STI Program. She made numerous corrections to grammar, spelling, formatting, and vocabulary. She also ensured that all figures and equations were included, keyed, and oriented accurately. I offered a few preliminary corrections that appeared in the original translation, such as the rendition of the word *Kunstsatz* in a fireworks rocket as "bursting charge" rather than "man-made charge" and of *Stab* on the same rocket as "guide stick" rather than "brace." I later went through the translated text more extensively and compared it with the original German. In the process, I

---

35. W. David Compton and Charles D. Benson, *Living and Working in Space: A History of Skylab* (Washington, D.C.: NASA SP-4208, 1983), esp. pp. 247-338, 381-386.
36. See for example, *The Aeronautics and Space Report of the President, Fiscal Year 1992 Activities* (Washington D.C.: Government Printing Office, 1993), pp. 11, 15-16.
37. Sylvia D. Fries, "2001 to 1994: Political Environment and the Design of NASA's Space Station System," *Technology and Culture* 29 (1988): 571.
38. See, e.g., the unpublished paper of Alex Roland, "The Evolution of Civil Space Station Concepts in the United States" (May 1983), seen in his biographical file, NASA Historical Reference Collection.
39. Hook, "Historical Review and Current Plans," pp. 8-10; *Aeronautics and Space Report, FY 1992*, Appendix C.

made numerous other changes to the original translation. For example, SCITRAN consistently translated the word *Betriebsstoff* as fuel when in many instances it clearly refers to both a fuel and an oxidizer. Hence, according to usage, I sometimes rendered it as "propellant" and in other instances left it as "fuel" where that translation seemed appropriate. Similarly, where appropriate I changed the translation of *Gestirn* from "star" to "planet" where the word clearly referred to one of the natural satellites of our sun. In general, I tried to ensure that the present translation followed American colloquial usage without deviating from the sense of the original German. I also added a number of asterisked footnotes to the text, largely to identify people Potočnik mentioned. (The numbered footnotes are Potočnik's own contributions.)

Following my editing, Dr. Ernst Stuhlinger kindly agreed to read the translation. Dr. Stuhlinger, who earned his Ph.D. in physics at the University of Tübingen in 1936 and whose involvement in rocketry and space work began at Peenemünde in 1943 and extended through service as associate director of science at the NASA-Marshall Space Flight Center through the end of 1975, has a much more intimate knowledge of the technological details about which Potočnik wrote than I do as well as a more subtle grasp of the nuances of Potočnik's Austrian German. He has painstakingly read through the original German and made numerous corrections to the translation that I missed. For example, the translators had rendered *fortgeschleudert* (literally, "flung away") as "accelerated away," whereas Dr. Stuhlinger improved that to the more colloquial and accurate "launched." Similarly, he changed the translation of *Schwerpunkt* from the correct dictionary meaning "center of mass" to the more exact "center of gravity." And where the translation rendered *liegender* in reference to a position of the human body as "lying," he changed that to the much more appropriate English word, "prone." In these and countless other ways, he made the translation both more accurate and more readable without altogether effacing the style of someone writing in Austrian German at a date well before many of today's technical terms had been coined.

As Dr. Stuhlinger wrote to me on 2 August 1994 after he had finished his final editing, "Noordung's way of writing is in his lovable and homely Austrian style, with many small words that do not contribute much to the content of a sentence, but rather to an easy flow of the language. Many of these little words," Stuhlinger went on, "can have a multitude of meanings, depending on the context; in looking up these words in the dictionary," it is easy to choose a wrong translation and thus change a sentence's meaning. "In numerous cases, I just struck out such words, because they are not really needed, and only burden the text; in other cases, I had to select another English translation." Despite such problems with translating the book, however, and despite "some basically incorrect views expressed by the author," Stuhlinger added "it is a remarkable book" that he thought should be made accessible to an audience not able to read it in the original Austrian German.

In my own editing and in writing this introduction, I have benefitted from the advice and assistance of a number of people. In listing them, I run the risk of forgetting some whose help I neglected to annotate in my notes, but the list should certainly include John Mulcahey, Jesco von Puttkamer, Otto Guess, Lee Saegesser, Shirley Campos, Beverly S. Lehrer, Jennifer Hopkins, Bill Skerrett, Adam Gruen, Timothy Cronen, Michael J. Neufeld, Howard E. McCurdy, and Roger D. Launius. Fred Ordway deserves special mention not only for writing the foreword but for sharing materials for an earlier translation he had planned in conjunction with Harry O. Ruppe, Willy Ley and others. In addition to doing the initial editing of TT-10002, Jennifer Garland worked with me closely in arranging for the translation, and I would like to thank her for being exceptionally cooperative and professional in that effort. Above all, however, Dr. Stuhlinger deserves credit for making the translation much more accurate than it would have been without his careful editing. With very minor changes, I have accepted his corrections of the text, but I alone retain responsibility for any errors that remain in the translation. Finally, let me thank Lynn Van der Veer and Steve Chambers for their extremely professional work designing the book and preparing it for publication.

[Noordung's] Introduction

Since the beginning of time, mankind has considered it as an expression of its Earthly weakness and inadequacy to be bound to the Earth, to be unable to free itself from the mysterious shackles of gravity. Not without good reason then has the concept of the transcendental always been associated with the idea of weightlessness, the power "to be able freely to rise into the sky." And most people even today still take it as a dogma that it is indeed unthinkable for Earthly beings ever to be able to escape the Earth. Is this point of view really justified?

Keep in mind: just a few decades ago, the belief indelibly impressed upon us was widespread that it is foolhardy to hope that we would ever be able to speed through the air like the birds. And today! In the face of this and similar superb proofs of the capability of science and technology, should mankind not dare now to tackle the last transportation problem for which a solution still eludes us: the problem of space travel? And logically: in the last few years, the "technical dream," which to date was only the stuff of fanciful novels, has become a "technical question" examined in the dispassionate works of scholars and engineers using all the support of mathematical, physical and technical knowledge and—has been deemed solvable.

The Power of Gravity

The most critical obstacle standing in the way of traveling in space is the gravitation of the Earth. Because a vehicle that is supposed to travel in outer space must be able not only to move; it must primarily and first of all move away from the Earth—i.e., against the force of gravity. It must be able to lift itself and its payload up many thousands, even hundreds of thousands of kilometers!

Because the force of gravity is an inertial force, we must first of all understand the other inertial forces existing in nature and, moreover, briefly examine what causes these forces, namely the two mechanical fundamental properties of mass; because the entire problem of space travel is based on these issues.

The first of these properties lies in the fact that all masses mutually attract (Law of Gravitation). The consequence of this phenomenon is that every mass exerts a so-called "force of mutual attraction" on every other mass. The attractive force that the celestial bodies exert on other masses by virtue of their total mass is called the force of gravity. The "force of gravity" exerted by the Earth is the reason that all objects on the Earth are "heavy", that is, they have more or less "weight" depending on whether they themselves have a larger or smaller mass. The force of mutual attraction (force of gravity) is then that much more significant, the greater the mass of the objects between which it acts. On the other hand, its strength decreases with increasing distance (more specifically, with the square of the latter), nevertheless without its effective range having a distinct boundary (Figure 1). Theoretically, the force becomes zero only at an infinite distance. Similar to the Earth, the sun, Moon and, for that matter, every celestial body exerts a force of gravity corresponding to its size.

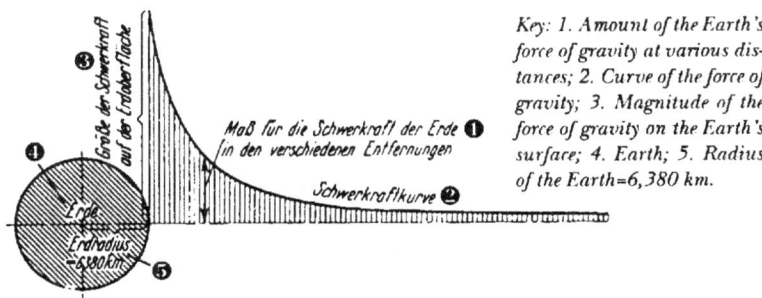

Key: 1. Amount of the Earth's force of gravity at various distances; 2. Curve of the force of gravity; 3. Magnitude of the force of gravity on the Earth's surface; 4. Earth; 5. Radius of the Earth=6,380 km.

Figure 1. The curve of the Earth's force of mutual attraction (force of gravity). The strength of the attraction, which decreases with the square of increasing distance is represented by the distance of the curve of the force of gravity from the horizontal axis.

The second fundamental property of mass lies in the fact that every mass is always striving to continue to remain in its current state of motion (Law of Inertia). Consequently, any mass whose motion is accelerated, decelerated or has its direction changed will resist this tendency by developing counteracting, so-called "forces of inertial mass" (Figure 2).

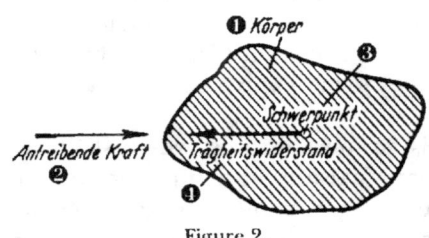

Figure 2.

Key:
1. Object; 2. Driving force; 3. Center of mass; 4. Inertia

In general, these are designated as inertia, or in a special case also as centrifugal force. The latter is the case when those forces occur due to the fact that mass is forced to move along a curved path. As is well known, the centrifugal force is always directed vertically outward from the curve of motion (Figure 3). All of these forces: force of gravity, inertia and the centrifugal force are inertial forces.

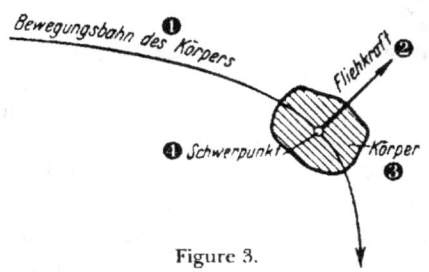

Figure 3.

Key: 1. Path of motion of the object; 2. Centrifugal force; 3. Object; 4. Center of gravity.

As mentioned previously, the effect of the Earth's force of gravity extends for an infinite distance, becoming weaker and weaker. We can consequently never completely escape the attractive range (the gravitational field) of the Earth, never reaching the actual gravitational boundary of the Earth. It can, nevertheless, be calculated what amount of work would theoretically be required in order to overcome the Earth's total gravitational field. To this end, an energy not less than 6,380 meter-tons would have to be used for every kilogram of load. Furthermore, it can be determined at what velocity an object would have to be launched from the Earth, so that it no longer returns to Earth. The velocity is 11,180 meters per second. This is the same velocity at which an object would strike the Earth's surface if it fell freely from an infinite distance onto the Earth. In order to impart this velocity to a kilogram of mass, the same amount of work of 6,380 meter-tons is required that would have to be expended to overcome the total Earth's gravitational field per kilogram of load.

If the Earth's attractive range could never actually be escaped, possibilities would nevertheless exist for an object to escape from the gravitational effect of the Earth, and more specifically, by the fact that it is also subjected to the effect of other inertial forces counterbalancing the Earth's force of gravity. According to our previous consideration about the fundamental properties of mass, only the following forces are possible: either the forces of mutual attraction of neighboring heavenly bodies or forces of inertial mass self-activated in the body in question.

The Practical Gravitational Boundary of the Earth

First of all, we want to examine the previously cited possibility. Because like the Earth every other celestial body also has a gravitational field that extends out indefinitely, losing more and more strength the further out it goes, we are—theoretically, at least—always under the simultaneous gravitational effect of all heavenly bodies. Of this effect, only the gravitational effect of the Earth and, to some degree, that of our Moon is noticeable to us, however. In the region close to the Earth's surface, in which mankind lives, the force of the Earth's attraction is so predominately overwhelming that the gravitational effect exerted by other celestial bodies for all practical purposes disappears compared to the Earth's attraction.

Something else happens, however, as soon as we distance ourselves from the Earth. Its attractive force continually decreases in its effect, while, on the other hand, the effect of the neighboring heavenly bodies

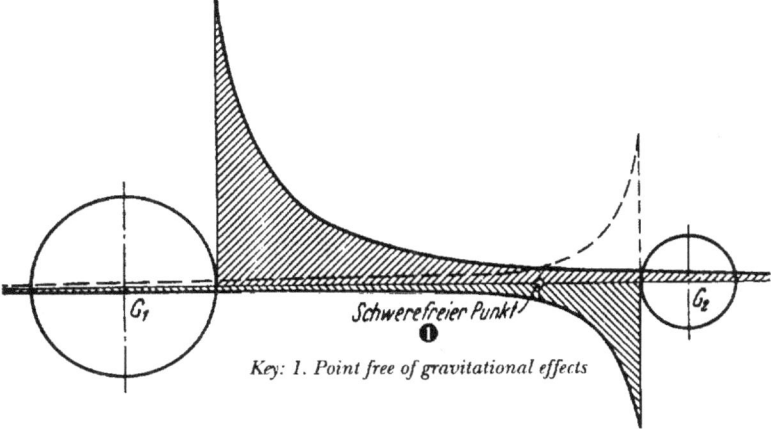

Key: 1. Point free of gravitational effects

Figure 4. The curve of the gravitational fields of two neighboring heavenly bodies G_1 and G_2 is represented as in Figure 1, with the exception that the gravitational curve of the smaller celestial body G_2 was drawn below the line connecting the centers because its attractive force counteracts that of the larger entity G_1. The point free of gravitational effects is located where both gravitational fields are opposite and equal to one another and, therefore, offset their effects.

increases continually. Since the effect counterbalances the Earth's force of gravity, a point must exist—seen from the Earth in every direction—at which these attractive forces maintain equilibrium concerning their strengths. On this side of that location, the gravitational effect of the Earth starts to dominate, while on the other side, that of the neighboring planet becomes greater. This can be designated as a practical boundary of the gravitational field of the Earth, a concept, however, that may not be interpreted in the strict sense, taking into consideration the large difference and continual changing of the position of the neighboring planets in relation to the Earth.

At individual points on the practical gravitational boundary (in general, on those that are on the straight line connecting the Earth and a neighboring planet), the attractive forces cancel one another according to the direction, such that at those points a completely weightless state exists. A point of this nature in outer space is designated as a so-called "point free of gravitational effects" (Figure 4). However, we would find ourselves at that point in an only insecure, unstable state of weightlessness, because at the slightest movement towards one side or the other, we are threatened with a plunge either onto the Earth or onto the neighboring planet.

Free Orbit

In order to attain a secure, stable state of weightlessness, we would have to escape the effect of gravity in the second way: with the aid of inertial forces.

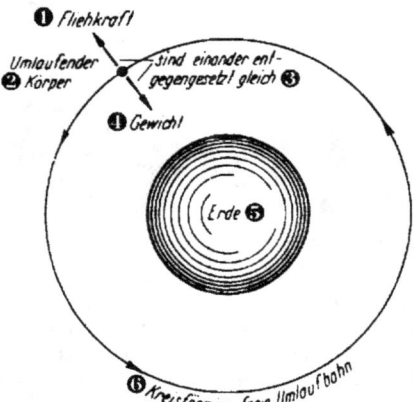

Figure 5. Circular free orbiting of an object around the Earth. The object's weight is offset by the centrifugal force generated during the orbiting. The object is, therefore, in a stable state of free suspension in relation to the Earth.

Key: 1. Centrifugal force; 2. Orbiting object; 3. Weight; 4. Are opposite and equal to one another; 5. Earth; 6. Circular free orbit

This is attained when the attracting celestial body, for example, the Earth, is orbiting in a free orbit at a corresponding velocity (gravitational motion). The centrifugal force occurring during the orbit and always directed outward maintains equilibrium with the attractive force—indeed, it is the only force when the motion is circular (Figure 5)—or simultaneously with other inertial forces occurring when the orbit has another form (ellipse, hyperbola, parabola, Figure 6).

All Moon and planet movements occur in a similar fashion. Because, by way of example, our Moon continuously orbits the Earth at an average velocity of approximately 1,000 meters per second, it does not fall onto the Earth even though it is in the Earth's range of attraction, but instead is suspended freely above it. And likewise the Earth does not plunge into the sun's molten sea for the simple reason that it continuously orbits the sun at an average velocity of approximately 30,000 meters per second. As a result of the centrifugal force generated during the orbit, the effect of the sun's gravity on the Earth is offset and, therefore, we perceive nothing of its existence. Compared to the sun, we are "weightless" in a "stable state of suspension;" from a practical point of view, we have been "removed from its gravitational effect."

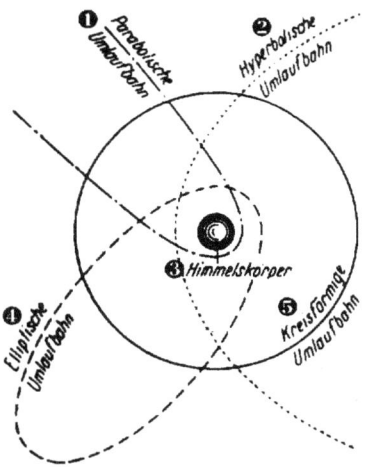

Figure 6. Various free orbits around a celestial body. In accordance with the laws of gravitational movement, a focal point of the orbit (the center in the case of a circle) must always coincide with the center of mass (center of gravity) of the orbiting celestial body.

Key: 1. Parabolic orbit; 2. Hyperbolic orbit; 3. Celestial body; 4. Elliptical orbit; 5. Circular orbit

The shorter the distance from the attracting celestial body in which this orbiting occurs, the stronger the effect of the attractive force at that point. Because of this, the counteracting centrifugal force and consequently the orbiting velocity must be correspondingly greater (because the centrifugal force increases with the square of the orbiting velocity). While, by way of example, an orbiting velocity of only about 1,000 meters per second suffices at a distance of the Moon from the Earth, this velocity would have to attain the value of approximately 8,000 meters per second for an object that is supposed to orbit near the Earth's surface in a suspended state (Figure 7).

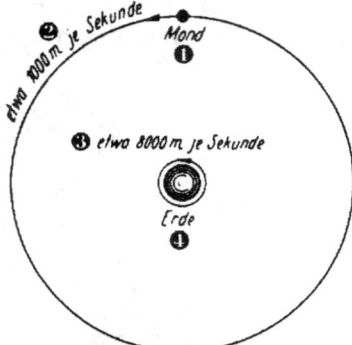

Figure 7. The orbiting velocity is that much greater the closer the free orbit movement occurs to the center of attraction.

Key: *1. Moon; 2. Approximately 1,000 meters per second; 3. Approximately 8,000 meters per second; 4. Earth*

In order to impart this velocity to an object, that is, to bring it into a stable state of suspension in relation to the Earth in such a manner, and as a result to free it from the Earth's gravity, an amount of work of about 3,200 meter-tons per kilogram of weight is required.

Maneuvering in the Gravitational Fields of Outer Space

Two basic possibilities exist in order to escape the gravitational effect of the Earth or of another heavenly body: reaching the practical gravitational boundary or transitioning into a free orbit. Which possibility will be employed depends on the intended goals.

Thus, for example, in the case of long-distance travel through outer space, it would generally depend on maneuvering in such a fashion that those celestial bodies, in whose range of attraction (gravitational field) the trip takes place, will be circled in a free orbit suspended in space (that is, only in suspension without power by a man-made force) if there is no intention to land on them. A longer trip would consist, however, of parts of orbits of this nature (suspension distances), with the transition from the gravitational field of one heavenly body into that of a neighboring one being caused generally by power from a man-made force.

If we want to remain at any desired altitude above a celestial body (e.g., the Earth) for a longer period, then we will continuously orbit that body at an appropriate velocity in a free circular orbit, if possible, and, therefore, remain over it in a stable state of suspension.

When ascending from the Earth or from another planet, we must finally strive either to attain the practical gravitational boundary and, as a result, the "total separation" (when foregoing a stable state of suspension) or transitioning into a free orbit and as a result into the "stable state of suspension" (when foregoing a total separation). Or, finally, we do not intend for the vehicle continually to escape the gravitational effect when ascending at all, but are satisfied to raise it to a certain altitude and to allow it to return immediately to Earth again after reaching this altitude (ballistic trajectory).

In reality, these differing cases will naturally not always be rigorously separated from one another, but frequently supplement one another. The ascent, however, will always have to take place by power from a manmade force and require a significant expenditure of energy, which—in the case when an ascending object is also to escape from the gravitational effect—for the Earth represents the enormous value of around 3,200 up to 6,400 meter-tons per kilogram of the load to be raised. Or—which amounts to the same thing—it requires imparting the huge, indeed cosmic velocity of approximately 8,000 to 11,200 meters per second, that is about 12 times the velocity of an artillery projectile!

The Armor Barrier of the Earth's Atmosphere

Besides the force of gravity, the atmosphere, which many celestial bodies have—naturally that of the Earth, in particular—also plays an extremely important role for space travel. While the atmosphere is very valuable for the landing, it, on the other hand, forms a particularly significant obstacle for the ascent.

According to observations of falling meteors and the northern lights phenomena, the height of the entire atmosphere of the Earth is estimated at several hundred (perhaps 400) kilometers (Figure 8). Nevertheless, only in its deepest layers several kilometers above the Earth, only on the "bottom of the sea of air" so to speak, does the air density exist that is necessary for the existence of life on Earth. For the air density decreases very quickly with increasing altitude and is, by way of example, one-half as great at an altitude of 5 km and only one-sixth as great at an altitude of 15 km as it is at sea level (Figure 9).

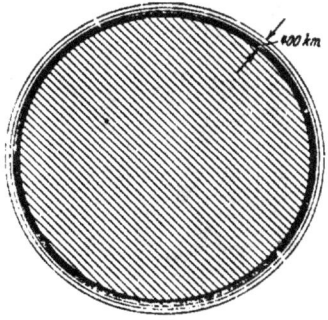

Figure 8. Assuming that the atmosphere is approximately 400 km high, the diagram shows its correct ratio to the Earth.

This condition is of critical importance for the question of space travel and is beneficial to it because, as is well known, air resists every moving object. During an increasing velocity of motion, the resistance increases, however, very rapidly, and more specifically, in a quadratic relationship. Within the dense air layers near the Earth, it reaches such high values at the extreme velocities considered for space travel that as a result the amount of work necessary for overcoming the gravitational field during ascent, as mentioned previously, is increased considerably and must also be taken into consideration to a substantial degree when building the vehicle. However, since the density of the air fortunately decreases rapidly with increasing altitude, its resistance also becomes smaller very quickly and can as a result be maintained within acceptable limits. Nevertheless, the atmosphere is a powerful obstacle during ascent for space travel. It virtually forms an armored shield surrounding the Earth on all sides. Later, we will get to know its importance for returning to Earth.

The Highest Altitudes Reached to Date

There has been no lack of attempts to reach the highest altitudes. Up to the present, mankind has been able to reach an altitude of 11,800 meters in an airplane, 12,000 meters in a free balloon, and 8,600 meters

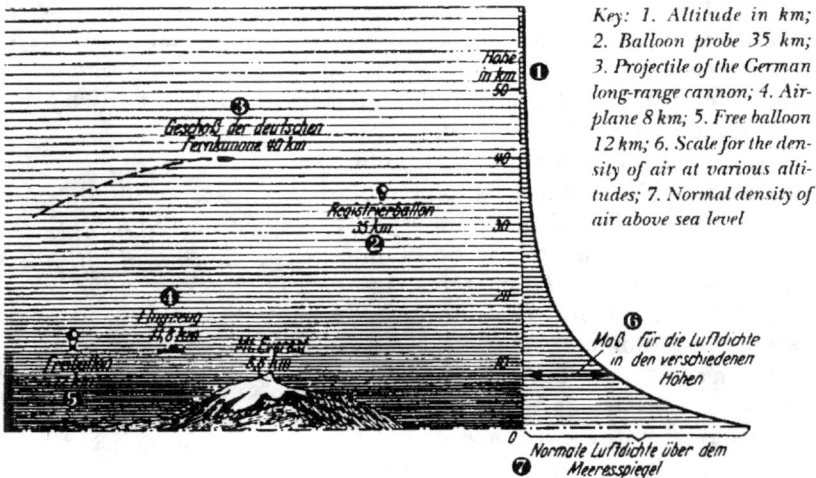

Key: 1. Altitude in km; 2. Balloon probe 35 km; 3. Projectile of the German long-range cannon; 4. Airplane 8 km; 5. Free balloon 12 km; 6. Scale for the density of air at various altitudes; 7. Normal density of air above sea level

Figure 9. With increasing altitude, the density of air decreases extremely rapidly, as can be seen from the curve drawn on the right and from the intensity of the shading.

on Mount Everest (Figure 9). So-called balloon probes have attained even higher altitudes. They are unmanned rubber balloons that are supposed to carry very lightweight recording devices as high as possible. Since the air pressure decreases continually with increasing altitude, the balloon expands more and more during the ascent until it finally bursts. The recording devices attached to a parachute gradually fall, recording automatically pressure, temperature and the humidity of the air. Balloon probes of this type have been able to reach an altitude up to approximately 35 kilometers. Moreover, the projectiles of the famous German long-range cannon, which fired on Paris, reached an altitude of approximately 40 kilometers. Nevertheless, what is all of this in comparison to the tremendous altitudes to which we would have to ascend in order to reach into empty outer space or even to distant celestial bodies!

The Cannon Shot into Outer Space

It appears obvious when searching for the means to escape the shackles of the Earth to think of firing from a correspondingly powerful giant cannon. This method would have to impart to the projectile the enormous energy that it requires for overcoming gravity and for going beyond the atmosphere as a kinetic force, that is, in the form of velocity. This requires, however, that the projectile must have already attained a velocity of not less than around 12,000 meters per second when leaving the ground if, besides the lifting energy, the energy for overcoming air drag is also taken into account.

Even if the means of present day technology would allow a giant cannon of this type to be built and to dare firing its projectile into space (as Professor H. Lorenz in Danzig has verified, we in reality do not currently have a propellant that would be sufficiently powerful for this purpose)—the result of this effort would not compensate for the enormous amounts of money required to this end.* At best, such an "ultra artillerist" would be able to boast about being the first one to accelerate an object from the Earth successfully or perhaps to have also fired at the Moon. Hardly anything more is gained by this because everything, payload, recording devices, or

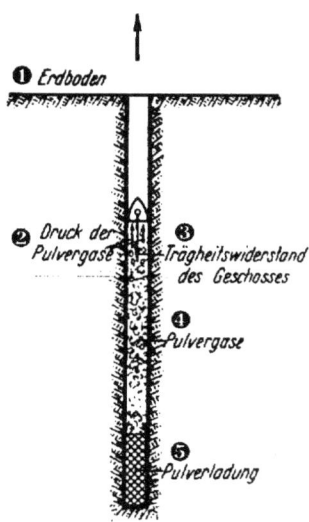

Figure 10. The Jules Verne giant cannon for firing at the Moon. The projectile is hollow and is designed for transporting human beings. The tube is embedded as a shaft in the ground.

Key: 1. Ground; 2. Pressure of the powder gases; 3. Inertia of the projectile; 4. Powder gases; 5. Powder charge

---

* Professor Hans Lorenz (1865-1940)—an expert on refrigeration, materials testing, and ballistics who taught at the technical institute in Danzig at this time, published in the *Zeitschrift des Vereins Deutscher Ingenieure* (periodical of the association of German engineers, 71, #19 [May 1927]: 651-654, 1128) a refutation of Hermann Oberth's calculations for space travel (see below). Oberth was able to demonstrate in *Die Rakete* (the rocket, the periodical of the Society for Spaceship Travel, vol. 1, #11 [Nov. 15, 1927]: 144-152 and #12 [Dec. 15, 1927]: 162-166; vol. 2, #6 [June 1928], 82-89) that Lorenz's calculations were false.

even passengers taken in this "projectile vehicle" during the trip, would be transformed into mush in the first second, because no doubt only solid steel would be able to withstand the enormous inertial pressure acting upon all parts of the projectile during the time of the firing, during which the projectile must be accelerated out of a state of rest to a velocity of 12,000 meters per second within a period of only a few seconds (Figure 10), completely ignoring the great heat occurring as a result of friction in the barrel of the cannon and especially in the atmosphere to be penetrated.

The Reactive Force

This method is, for all practical purposes, not usable. The energy that the space vehicle requires for overcoming gravity and air drag, as well as for moving in empty space, must be supplied to it in another manner, that is, by way of example, bound in the propellants carried on board the vehicle during the trip. Furthermore, a propulsion motor must also be available that allows the propulsion force during the flight to change or even shut off, to alter the direction of flight, and to work its way up gradually to those high, almost cosmic velocities necessary for space flight without endangering passengers or the payload.

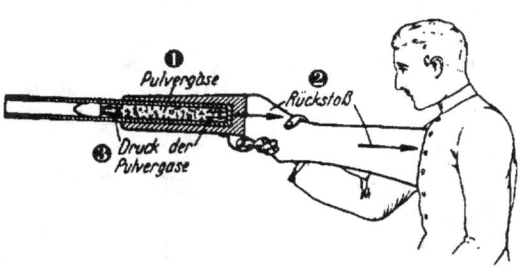

Figure 11. The "reactive force" or recoil when firing a rifle.

Key: 1. Powder gases; 2. Recoil; 3. Pressure of the powder gases.

But how do we achieve all of this? How is movement supposed to be possible in the first place since in empty space neither air nor other objects are available on which the vehicle can support itself (or push off from, in a manner of speaking) in order to continue its movement in accordance with one of the methods used to date? (Movement by foot for animals and human beings, flapping of wings by birds, driving wheels for rolling trucks, screws of ships, propellers, etc.)

A generally known physical phenomenon offers the means for this. Whoever has fired a powerful rifle (and in the present generation, these people ought not to be in short supply) has, no doubt, clearly felt the so-called "recoil" (maybe the experience was not altogether a pleasant one). This is a powerful action that the rifle transfers to the shooter against the direction of discharge when firing. As a result, the powder gases also press back onto the rifle with the same force at which they drive the projectile forward and, therefore, attempt to move the rifle backwards (Figure 11).

However, even in daily life, the reaction phenomenon can be observed again and again, although generally not in such a total sense: thus, for example, when a movable object is pushed away with the hand (Figure 12), exactly the same thrust then imparted to the object is, as is well known, also received by us at the same time in an opposite direction as a matter of course. Stated more precisely: this "reaction" is that much stronger, and we will as a result be pushed back that much further, the harder we pushed. However, the "velocity of repulsion," which the affected object being pushed away attains as a result, is also that much greater. On the other hand, we will be able to impart a velocity that much greater to the object being pushed away with one and the same force, the less weight the object has (i.e.,

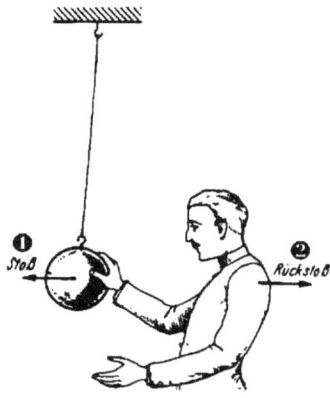

Figure 12. Even when a person quickly shoves an easily movable, bulkier object (e.g., a freely suspended iron ball) away from himself, he receives a noticeable "reactive force" automatically.

Key: 1. Action; 2. Reaction.

the smaller the mass). And likewise we will also fall back that much further, the lighter we are (and the less we will fall back, the heavier we are).

The physical law that applies to this phenomenon is called the "law maintaining the center of gravity." It states that the common center of gravity of a system of objects always remains at rest if they are set in motion only by internal forces, i.e., only by forces acting among these objects.

In our first example, the pressure of powder gases is the internal force acting between the two objects:

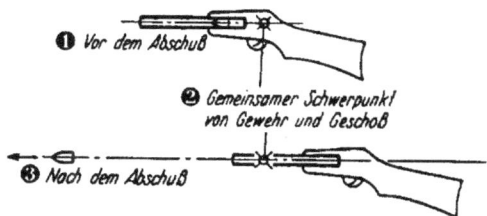

Figure 13. If the "reaction" of the rifle is not absorbed, it continually moves backwards (after firing), and more specifically, in such a manner that the common center of gravity of rifle and projectile remains at rest.

Key: 1. Prior to firing; 2. Common center of gravity of the rifle and projectile; 3. After firing

projectile and rifle. While under its influence the very small projectile receives a velocity of many hundreds of meters per second, the velocity, on the other hand, that the much heavier rifle attains in an opposite direction is so small that the resulting recoil can be absorbed by the shooter with his shoulder. If the person firing the rifle did not absorb the recoil and permitted the rifle to move backwards unrestrictedly (Figure 13), then the common center of gravity of the projectile and rifle would actually remain at rest (at the point where it was before firing), and the rifle would now be moving backwards.

The Reaction Vehicle

If the rifle was now attached to a light-weight cart (Figure 14) and fired, it would be set in motion by the force of the recoil. If the rifle was fired continually and rapidly, approximately similar to a machine gun, then the cart would be accelerated, and could also climb, etc. This would be a vehicle with reaction propulsion, not the most perfect, however. The continual movement of a vehicle of this type takes place as a result of the fact that it continually accelerates parts of its own mass (the projectiles in the previous example) opposite to the direction of motion and is repelled by these accelerated parts of mass.

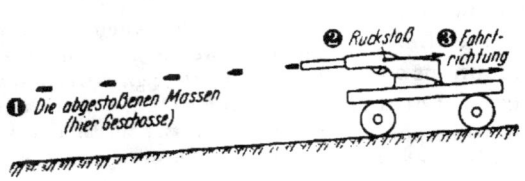

Figure 14. A primitive vehicle with reaction propulsion: The cart is moved by continuous firing of a rifle, as a result of the "reaction" generated thereby.

Key: 1. The masses flung away (the projectiles in this case); 2. Recoil; 3. Direction of travel

It is clear as a result that this type of propulsion will then be useful when the vehicle is in empty space and its environment has neither air nor something else available by which a repulsion would be possible. Indeed, the propulsion by recoil will only then be able to develop its greatest efficiency because all external resistances disappear.

For the engineering design of a vehicle of this type, we must now strive to ensure that for generating a specific propulsive force, on the one hand, as little mass as possible must be expelled and, on the other hand, that its expulsion proceeds in as simple and operationally safe a way, as possible.

To satisfy the first requirement, it is basically necessary that the velocity of expulsion be as large as possible. In accordance with what has already been stated, this can be easily understood even without mathematical substantiation, solely through intuition: for the greater the velocity with which I push an object away from me, the greater the force I have to apply against it; in accordance with what has already been stated, then the greater the opposite force will be that reacts on me as a result; this is the reaction produced by the expulsion of precisely this mass.

Furthermore, it is not necessary that larger parts of mass are expelled over longer time intervals, but rather that masses as small as possible are expelled in an uninterrupted sequence. Why this also contributes to keeping the masses to be expelled as low as possible, follows from mathematical studies that will not be used here, however. As can be easily understood, the latter must be required in the interest of operational safety, because the propulsive thrust would otherwise occur in jolts, something that would be damaging to the vehicle and its contents. Only a propulsive force acting as smoothly as possible is useful from a practical standpoint.

The Rocket

These conditions can best be met when the expulsion of the masses is obtained by burning suitable substances carried on the vehicle and by permitting the resulting gases of combustion to escape towards the rear—"to exhaust." In this manner, the masses are expelled in the smallest particles (molecules of the combustion gases), and the energy being freed during the combustion and being converted into gas pressure provides the necessary "internal power" for this process.

The well known fireworks rocket represents a vehicle of this type in a simple implementation (Figure 15). Its purpose is to lift a so-called "bursting charge": there are all sorts of fireworks that explode after reaching a certain altitude either to please the eye in a spectacular shower of sparks or (in warfare, by way of example) to provide for lighting and signaling.

The continual movement (lifting) of a fireworks rocket of this type takes place as a result of a powder charge carried in the rocket, desig-

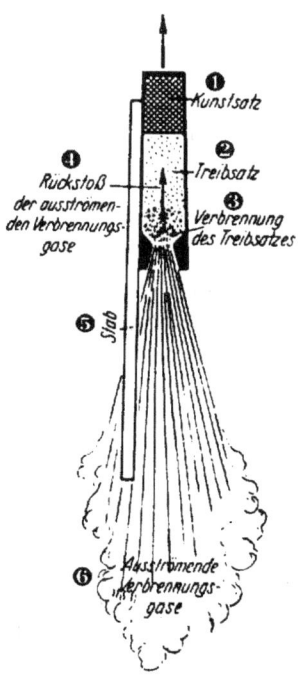

Figure 15. Fireworks rocket in a longitudinal section. The attached guide stick serves to inhibit tumbling of the rocket.

Key: 1. Bursting charge; 2. Propellant; 3. Combustion of the propellant; 4. Reaction of the escaping combustion gases; 5. Guide stick; 6. Escaping combustion gases

nated as the "propellant." It is ignited when the rocket takes off and then gradually burns out during the climb, with the resulting combustion gases escaping towards the rear (downward) and as a result—by virtue of its reaction effect—producing a continuous propulsion force directed forward (up) in the same way as was previously discussed.

However, a rocket that is supposed to serve as a vehicle for outer space would, to be sure, have to look considerably different from a simple fireworks rocket.

Previous Researchers Addressing the Problem of Space Flight

The idea that the reaction principle is suitable for the propulsion of space vehicles is not new. Around 1660, the Frenchman Cyrano de Bergerac in his novels described, to be sure in a very fantastic way, space travels in vehicles lifted by rockets. Not much later, the famous English scholar Isaac Newton pointed out in a scientific form the possibilities of being able to move forward even in a vacuum using the reaction process. In 1841, the Englishman Charles Golightly registered a patent for a rocket flight machine. Around 1890, the German Hermann Ganswindt and a few years later the Russian Tsiolkovsky made similar suggestions public for the first time. Similarly, the famous French author Jules Verne discussed in one of his writings the application of rockets for purposes of propulsion, although only in passing. The idea of a space ship powered by the effects of rockets emerged, however, very definitely in a novel by the German physicist Kurt Laßwitz.*

---

* Savinien Cyrano de Bergerac (1619-1655) was a French satirist and dramatist whose works included *A Voyage to the Moon*, published posthumously in 1656 in the original French. Isaac Newton (1642-1727), of course, was the famous English physicist and mathematician who formulated the three fundamental laws of motion, including the third law that applies to rocket operation in the vacuum of space. Charles Golightly registered a patent in England in 1841 for "motive power," but it contained no specifications. It is not clear that it was a serious patent, and in any event, Golightly is a historical mystery about whom virtually nothing is known. A Mr. Golightly was earlier the subject of cartoons showing him riding on a steam-powered rocket, however. Hermann Ganswindt (1856-1934), a German inventor with contradictory ideas about rocket power, nevertheless preceded Tsiolkovsky (see next footnote) in publicizing ideas about space flight as early as 1881. Jules Verne (1828-1905) was the French writer whose books pretty much established science fiction as a genre of fiction. Among his novels were *From the Earth to the Moon* (published in French in 1865), which greatly inspired the pioneers of space flight. Kurt (also spelled Kurd) Laßwitz (1848-1910) wrote the novel *On Two Planets*, published in German in 1897, in which Martians came to Earth in a space ship that nullified gravity.

Yet only in the most recent times, have serious scientific advances been undertaken in this discipline, and indeed apparently from many sides at the same time: a relevant work by Professor Dr. Robert H. Goddard appeared in 1919. The work of Professor Hermann Oberth, a Transylvanian Saxon, followed in 1923. A popular representation by Max Valier, an author from Munich, was produced in 1924, and a study by Dr. Walter Hohmann, an engineer from Essen, in 1925. Publications by Dr. Franz Edler von Hoefft, a chemist from Vienna, followed in 1926. New relevant writings by Tsiolkovsky, a Russian professor, were published in 1925 and 1927.*

Also, several novels, which treated the space flight problem by building on the results of the most recent scientific research specified above, have appeared in the last few years, in particular, those from Otto Willi Gail standing out.**

Before we turn our attention now to the discussion of the various recommendations known to date, something first must be said regarding the fundamentals of the technology of motion and of the structure of rocket space vehicles.

The Travel Velocity and the Efficiency of Rocket Vehicles

It is very important and characteristic of the reaction vehicle that the travel velocity may not be selected arbitrarily, but is already specified in general due to the special type of its propulsion. Since continual motion of a vehicle of this nature occurs as a result of the fact that it expels parts of its own mass, this phenomenon must be regulated in such a manner

---

* Konstantin E. Tsiolkovsky (1857-1935), Robert H. Goddard (1882-1945), and Hermann Oberth (1894-1989) are generally recognized as the three preeminent fathers of spaceflight. Tsiolkovsky and Oberth, Russian and Rumanian-German by nationality, were primarily theorists, whereas the American Professor Goddard not only wrote about rocket theory but engaged in rocket development and testing. The Austrian Max Valier (1895-1930) was a popularizer of Oberth's ideas who died in an explosion of a rocket motor mounted in a car. Walter Hohmann (1880-1945), actually an architect in Essen, published *The Attainability of Celestial Bodies* in German in 1925, in which he dealt with theoretical problems of space travel and discussed what came to be called the Hohmann orbits whereby spacecraft follow elliptical orbits tangent to the paths of the Earth and the target planet to reach the latter with the least expenditure of fuel and energy. Dr. Franz von Hoefft (1882-1954) was the principal founder of the Austrian Society for High-Altitude Research. He wrote essays on propellants for rockets and "From Aviation to Space Travel" in *The Possibility of Space Travel*, edited by Willy Ley, that appeared in German in 1928.

** Otto Willi Gail (1896-1956) was a German science fiction writer who wrote such novels as *The Shot into Infinity* (published in German in 1925) and *The Shot from the Moon* (1926).

that all masses have, if possible, released their total energy to the vehicle following a successful expulsion, because the portion of energy the masses retain is irrevocably lost. As is well known, energy of this type constitutes the kinetic force inherent in every object in motion. If now no more energy is supposed to be available in the expelling masses, then they must be at rest vis-a-vis the environment (better stated: vis-a-vis their state of motion before starting) following expulsion. In order, however, to achieve this, the travel velocity must be of the same magnitude as the velocity of expulsion, because the velocity, which the masses have before their expulsion (that is, still as parts of the vehicle), is just offset by the velocity that was imparted to them in an opposite direction during the expulsion (Figure 16). As a result of the expulsion, the masses subsequently arrive in a relative state of rest and drop vertically to the ground as free falling objects.

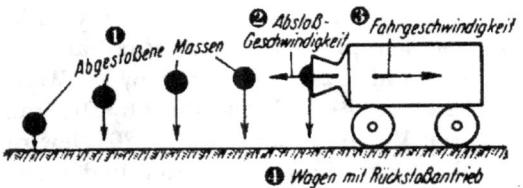

Figure 16. The travel velocity is equal to the velocity of expulsion. Consequently, the velocity of the expelled masses equals zero after the expulsion, as can be seen from the figure by the fact that they drop vertically.

Key: 1. Expelled masses; 2. Velocity of expulsion; 3. Travel velocity; 4. Cart with reactive propulsion

Under this assumption in the reaction process, no energy is lost; reaction itself works with a (mechanical) efficiency of 100 percent (Figure 16). If the travel velocity was, on the other hand, smaller or larger than the velocity of expulsion, then this "efficiency of reactive propulsion" would also be correspondingly low (Figure 17). It is completely zero as soon as the vehicle

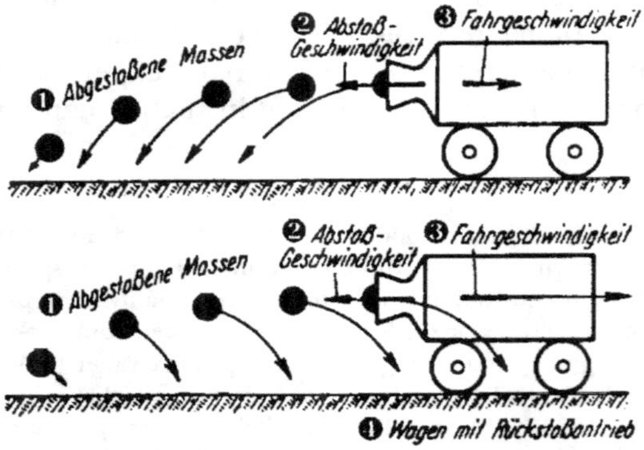

Figure 17. The travel velocity is smaller (top diagram) or larger (lower diagram) than the velocity of expulsion. The expelled masses still have, therefore, a portion of their velocity of expulsion (top diagram) or their travel velocity (lower diagram) following expulsion, with the masses sloping as they fall to the ground, as can be seen in the figure.

Key: 1. Expelled masses; 2. Velocity of expulsion; 3. Travel velocity; 4. Cart with reactive propulsion comes to rest during an operating propulsion.

This can be mathematically verified in a simple manner, something we want to do here by taking into consideration the critical importance of the question of efficiency for the rocket vehicle. If the general expression for efficiency is employed in the present case: "Ratio of the energy gained to the energy expended"[1], then the following formula is arrived at

$$\eta r = \left(2 - \frac{v}{c}\right)\frac{v}{c}$$

as an expression for the efficiency of the reaction η_r as a function of the instantaneous ratio between travel velocity v and the velocity of repulsion c.

In Table 1, the efficiency of the reaction η_r is computed for various values of this $\frac{v}{c}$ ratio using the above formula. If, for example, the $\frac{v}{c}$ ratio was equal to 0.1 (i.e., v=0.1 c, thus the travel velocity is only one-tenth as large as the velocity of expulsion), then the efficiency of the reaction would only be 19 percent. For $\frac{v}{c}=0.5$ (when the travel velocity

---

1. η_r = Energy gained/Energy expended

    = (Energy expended - Energy lost) / Energy expended

Energy expended = $\frac{mc^2}{2}$,

Energy lost = $\frac{m(c-v)^2}{2}$,

with m being the observed repulsion masses and (c–v) being their speed of motion still remaining after the repulsion (according to what has already be stated, this means a kinetic force lost to the vehicle).

It follows from that:

$$\eta r = \frac{\frac{mc^2}{2} - \frac{m(c-v)^2}{2}}{\frac{mc^2}{2}} = \left(2 - \frac{v}{c}\right)\frac{v}{c}$$

Table 1

Ratio of the travel velocity v to the velocity of expulsion c $\frac{v}{c}$	Efficiency of the Reaction η_r	
	$\eta_r = \left(2 - \frac{v}{c}\right)\frac{v}{c}$	η_r in percentages (rounded-up)
0	0	0
0.01	0.0199	2
0.05	0.0975	10
0.1	0.19	19
0.2	0.36	36
0.5	0.75	75
0.8	0.96	96
1	1	**100**
1.2	0.96	96
1.5	0.75	75
1.8	0.36	36
2	0	0
2.5	-1.25	-125
3	-3	-300
4	-8	-800
5	-15	-1500

is one-half as large as the velocity of repulsion), the efficiency would be 75 percent, and for $\frac{v}{c}=1$ (the travel velocity equals the velocity of expulsion)—in agreement with our previous consideration—the efficiency would even be 100 percent. If the $\frac{v}{c}$ ratio becomes greater than 1 (the travel velocity exceeds the velocity of expulsion), the efficiency of the reaction is diminished again and, finally, for $\frac{v}{c}=2$ it again goes through zero and even becomes negative (at travel velocities more than twice as large as the velocity of expulsion).

The latter appears paradoxical at first glance because the vehicle gains a travel velocity as a result of expulsion and apparently gains a kinetic force as a result! Since the loss of energy, resulting through the separation of the expulsion mass loaded very heavily with a kinetic force due to the large travel velocity, now exceeds the energy gain realized by the expulsion, an energy loss nevertheless results for the vehicle from the entire process—despite the velocity increase of the vehicle caused as a result. The energy loss is expressed mathematically by the negative sign of the efficiency. Nonetheless, these efficiencies resulting for large values of the $\frac{v}{c}$ ratio have, in reality, only a more or less theoretical value.

It can, however, clearly and distinctly be seen from the table how advantageous and, therefore, important it is that the travel velocity approaches as much as possible that of the expulsion in order to achieve a good efficiency of reaction, but slight differences (even up to $v=0.5$ c and/or $v=1.5$ c) are, nevertheless, not too important because fluctuations of the efficiency near its maximum are fairly slight. Accordingly, it can be stated that the optimum travel velocity of a rocket vehicle is approximately between one-half and one and one-half times its velocity of expulsion.

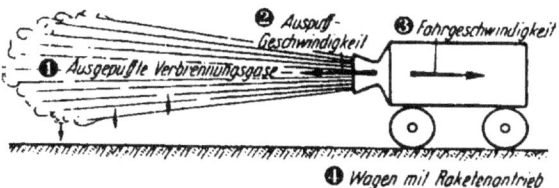

Figure 18. For a rocket vehicle, the travel velocity must as much as possible be equal to the exhaust velocity.

Key: 1. Exhausted gases of combustion; 2. Exhaust velocity; 3. Travel velocity; 4. Cart with rocket propulsion

When, as is the case here, the reaction vehicle is a rocket vehicle and consequently the expulsion of masses takes place through appropriate combustion and exhausting of propellants carried on the vehicle, then, in the sense of the requirement just identified, the travel velocity must be as much as possible of the same magnitude as the exhaust velocity (Figure 18). To a certain extent, this again requires, however, that the travel velocity conforms to the type of propellants used in each case, because each has its own maximum achievable exhaust velocity.

This fundamental requirement of rocket technology is above all now critical for the application of rocket vehicles. According to what has already been stated, the velocity of expulsion should then be as large as possible. Actually, the possible exhaust velocities are thousands of meters per second and, therefore, the travel velocity must likewise attain a correspondingly enormous high value that is not possible for all vehicles known to date, if the efficiency is supposed to have a level still usable in a practical application.

Table 2

1	2							3	
Travel velocity v in	Efficiency of the Reaction $$\eta r = \left(2 - \frac{v}{c}\right)\frac{v}{c}$$							Total Efficiency of the vehicle propulsion $\eta = \eta_r \eta_i$ for benzene and liquid oxygen as propellants	
	Expressed in percentages for the following velocities of expulsion c in m/sec:								
km/h | m/s	1000	2000	2500	3000	3500	4000	5000	2000	3500
40 | 11	2.2	1.2	0.9	0.7	0.6	0.5	0.4	0.2	0.4
100 | 28	4.6	2.8	2.2	1.8	1.6	1.4	1.2	0.6	1
200 | 56	11	5.5	4.5	3.8	3.2	2.8	2.2	1.1	2
300 | 83	16	8	6.5	5.5	4.7	4	3.4	1.6	3
500 | 140	26	13	11	9	8	7	5.5	2.7	5
700 | 200	36	19	15	13	11	10	8	4	7
1000 | 300	51	28	23	19	16	14	12	6	10
1800 | 500	75	44	36	31	27	23	19	9	17
3000 | 1000	100	75	64	56	50	44	36	15	31
5400 | 1500	75	94	84	75	67	60	51	19	42
7200 | 2000	0	100	96	89	81	75	64	20	50
9000 | 2500	-125	94	100	97	92	86	75	19	57
10800 | 3000	-300	75	96	100	98	94	84	15	61
12600 | 3500	-525	44	84	97	100	99	91	9	62
14400 | 4000	-800	0	64	89	98	100	96	0	61
18000 | 5000	-1500	-125	0	56	81	94	100	-25	50
21600 | 6000		-300	-96	0	50	75	96	-61	31
25200 | 7000		-520	-220	-77	0	44	70	-111	0
28800 | 8000		-800	-380	-175	-64	0	64	-160	-40
36000 | 10000		-1500	-800	-440	-250	-125	0	-300	-160
45000 | 12500			-1500	-900	-560	-350	-125		-350

This can be clearly seen from Table 2, in which the efficiencies corresponding to the travel velocities at various velocities of expulsion are determined for single important travel velocities (listed in Column 1). It can be seen from Column 2 of the table, which lists the efficiency of reaction, how uneconomical the rocket propulsion is at velocities (of at most several hundred kilometers per hour) attainable by our present vehicles.

This stands out much more drastically if, as expressed in Column 3, the total efficiency is considered. This is arrived at by taking into account the losses that are related to the generation of the velocity of expulsion (as a result of combustion and exhausting of the propellants). These losses have the effect that only an exhaust velocity smaller than the velocity that would be theoretically attainable in the best case for those propellants can ever be realized in practice. As will subsequently be discussed in detail,[2] the practical utilization of the propellants could probably be brought up to approximately 60 percent. For benzene by way of example, an exhaust velocity of 3,500 meters per second at 62 percent and one of 2,000 meters per second at 20 percent would result. Column 3 of Table 2 shows the total efficiency for both cases (the efficiency is now only 62 percent and/or 20 percent of the corresponding values in Column 2, in the sense of the statements made).

As can be seen from these values, the total efficiency—even for travel velocities of many hundreds of kilometers per hour—is still so low that, ignoring certain special purposes for which the question of economy is not important, a far-reaching practical application of rocket propulsion can hardly be considered for any of our customary means of ground transportation.

On the other hand, the situation becomes entirely different if very high travel velocities are taken into consideration. Even at supersonic speeds that are not excessively large, the efficiency is considerably better and attains even extremely favorable values at still higher, almost cosmic travel velocities in the range of thousands of meters per second (up to tens of thousands of kilometers per hour), as can be seen in Table 2.

It can, therefore, be interpreted as a particularly advantageous encounter of conditions that these high travel velocities are not only possible (no resistance to motion in empty space!) for space vehicles for which the reaction represents the only practical type of propulsion, but even represent an absolute necessity. How otherwise could those enormous distances of outer space be covered in acceptable human travel times? A danger, however, that excessively high velocities could perhaps cause harm does not exist, because we are not directly aware whatsoever of velocity per se, regardless of how high it may be. After all as "passen-

---

2. See pages 40-41 for a discussion of the "internal efficiency" of the rocket motor.

Table 3

Kilometers per hour km/hour	Meters per second m/sec	Kilometers per second km/sec
5	1.39	0.00139
10	2.78	0.00278
30	8.34	0.00834
50	13.9	0.0139
70	19.5	0.0195
90	25.0	0.0250
100	27.8	0.0278
150	41.7	0.0417
200	55.6	0.0556
300	83.4	0.0834
360	100	0.1
500	139	0.139
700	195	0.195
720	200	0.2
1000	278	0.278
1080	300	0.3
1190	330	0.33
1800	500	0.5
2000	556	0.556
2520	700	0.7
3000	834	0.834
3600	1000	1
5400	1500	1.5
7200	2000	2
9000	2500	2.5
10800	3000	3
12600	3500	3.5
14400	4000	4
18000	5000	5
21600	6000	6
25200	7000	7
28800	8000	8
36000	10000	10
40300	11180	11.18
45000	12500	12.5
54000	15000	15
72000	20000	20

gers of our Earth," we are continually racing through space in unswerving paths around the sun at a velocity of 30,000 meters per second, without experiencing the slightest effect. However, the "accelerations" resulting from forced velocity changes are a different matter altogether, as we will see later.

Table 3 permits a comparison to be made more easily among the various travel velocities under consideration here—something that is otherwise fairly difficult due to the difference of the customary systems of notation (kilometers per hour for present day vehicles, meters or kilometers per second for space travel).

The Ascent

Of the important components of space fight—the ascent, the long-distance travel through outer space, and the return to Earth (the land-

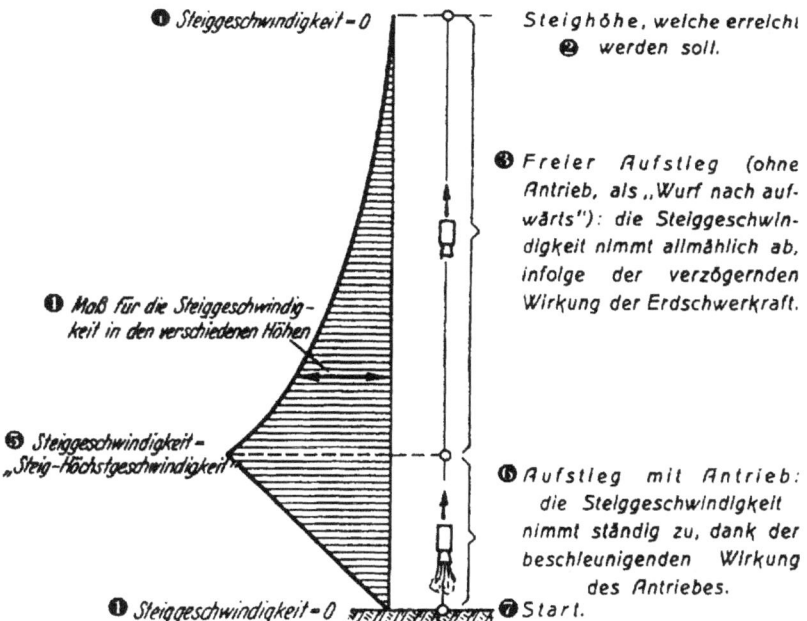

Figure 19. Vertical ascent—"steep ascent"—of a space rocket.

Key: 1. Climbing velocity=0; 2. Climbing altitude that is supposed to be reached; 3. Free ascent (without power as a "hurl upwards"): the climbing velocity decreases gradually as a result of the decelerating effect of the Earth's gravity; 4. Measure for the climbing velocity at various altitudes; 5. Climbing velocity="highest velocity of climbing"; 6. Power ascent: the climbing velocity increases continuously thanks to the accelerating effect of the propulsion system; 7. Launch.

ing)—we want to address only the most critical component at this point: the ascent. The ascent represents by far the greatest demands placed on the performance of the propulsion system and is also, therefore, of critical importance for the structure of the entire vehicle.

For implementing the ascent, two fundamental possibilities, the "steep ascent" and "flat ascent," present themselves as the ones mentioned at the beginning[3] in the section about movement in the gravity fields of outer space. In the case of the steep ascent, the vehicle is lifted in at least an approximately vertical direction. During the ascent, the climbing velocity, starting at zero, initially increases continuously thanks to the thrusting force of the reaction propulsion system (Figure 19); more specifically, it increases until a high climbing velocity is attained—we will designate it as the "maximum velocity of climbing"—such that now the power can be shut off and the continued ascent, as a "hurl upward," can continually proceed up to the desired altitude only under the effect of the kinetic energy that has been stored in the vehicle.

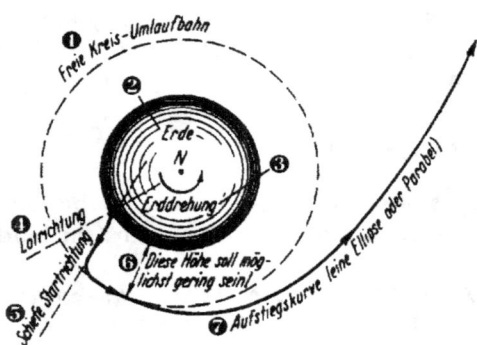

Figure 20. "Flat ascent" of a space rocket. The expenditure of energy for the ascent is the lowest in this case.

Key: 1. Free circular orbit; 2. Earth; 3. Earth rotation; 4. Vertical direction; 5. Inclined direction of launch; 6. This altitude should be as low as possible!; 7. Ascent curve (an ellipse or parabola)

In the case of the flat ascent, on the other hand, the vehicle is not lifted vertically, but in an inclined (sloped) direction, and it is a matter not so much of attaining an altitude but rather, more importantly, of gaining horizontal velocity and increasing it until the orbiting velocity necessary for free orbital motion and consequently the "stable state of suspension" are attained (Figures 5 and 20). We will examine this type of ascent in more detail later.

First, however, we want to examine some other points, including the question: How is efficiency varying during the ascent? For regardless how the ascent takes place, the required final velocity can only gradually be attained in any case, leading to the consequence that the travel (climbing) velocity of the space rocket will be lower in the beginning and greater

---

3. See pages 8-9.

later on (depending on the altitude of the final velocity) than the velocity of expulsion. Accordingly, the efficiency of the propulsion system must also be constantly changing during the power ascent, because the efficiency, in accordance with our previous definitions, is a function of the ratio of the values of the velocities of travel and expulsion (see Table 1, page 20). Accordingly in the beginning, it will only be low, increasing gradually with an increasing climbing velocity, and will finally exceed its maximum (if the final velocity to be attained is correspondingly large) and will then drop again.

In order to be able to visualize the magnitude of the efficiency under these conditions, the "average efficiency of the propulsion system" η_{rm} resulting during the duration of the propulsion must be taken into consideration. As can be easily seen, this efficiency is a function, on the one hand, of the velocity of expulsion c, which we want to assume as constant for the entire propulsion phase, and, on the other hand, of the final velocity v' attained at the end of the propulsion period.

The following formula provides an explanation on this point:

$$\eta_{rm} = \frac{\left(\frac{v'}{c}\right)^2}{e^{\frac{v'}{c}} - 1} \quad *),^4$$

4. The average efficiency of the reaction

$$\eta_{rm} = \frac{\text{Energy gained}}{\text{Energy expended}} =$$

$$\frac{\text{Kinetic force of the final mass M at the final velocity v'}}{\text{Kinetic force of the expelled mass } (M_0 - M) \text{ at the velocity of expulsion c:}}$$

$$\eta_{rm} = \frac{\frac{Mv'^2}{2}}{\frac{(M_0 - M)c^2}{2}}$$

With $M_0 = Me^{\frac{v'}{c}}$ the following results (see page 36):

$$\eta_{rm} = \frac{Mv'^2}{\left(Mc^{\frac{v'}{c}} - M\right)c^2} = \frac{\left(\frac{v'}{c}\right)^2}{e^{\frac{v'}{c}} - 1}$$

Table 4 was prepared using this formula.

Table 4

Ratio of the final velocity v' to the velocity of expulsion c: $\dfrac{v'}{c}$	Average efficiency of the propulsion system η_{rm} during the acceleration phase	
	$\eta_{rm} = \dfrac{\left(\dfrac{v'}{c}\right)^2}{e^{\frac{v'}{c}} - 1}$	η_{rm} in percentages
0	0	0
0.2	0.18	18
0.6	0.44	44
1	0.58	58
1.2	0.62	62
1.4	0.64	64
1.59	**0.65**	**65**
1.8	0.64	64
2	0.63	63
2.2	0.61	61
2.6	0.54	54
3	0.47	47
4	0.30	30
5	0.17	17
6	0.09	9
7	0.04	4

The table shows the average efficiency of the propulsion system as a function of the ratio of the final velocity v' attained at the end of the propulsion phase to the velocity of expulsion c existing during the propulsion phase, that is, a function of $\dfrac{v'}{c}$. Accordingly by way of example at a velocity of expulsion of c=3,000 meters per second and for a propulsion phase at the end of which the final velocity of v=3,000 meters per second is attained (that is, for $\dfrac{v'}{c}=1$), the average efficiency of the propulsion system would be 58 percent. It would be 30 percent for the final velocity of v=12,000 meters per second $\left(\text{that is, } \dfrac{v'}{c}=4\right)$, and so on. In the best case (that is, for $\dfrac{v'}{c}=1.59$,) in our example, the efficiency would even attain 65 percent for a propulsion phase at a final velocity of v'=4,770 meters per second.

In any case it can be seen that even during the ascent, the efficiency is generally still not unfavorable despite the fluctuations in the ratio of the velocities of travel and expulsion.

Besides the efficiency problem being of interest in all cases, a second issue of extreme importance exists especially for the ascent. As soon as the launch has taken place and, thus, the vehicle has lifted off its support (solid base or suspension, water-surface, launch balloon, etc.), it is carried only by the propulsion system (Figure 21), something—according to the nature of the reactive force—that depends on to a continual expenditure of energy (fuel consumption). As a result, that amount of propellants required for the lift-off is increased by a further, not insignificant value. This condition lasts only until—depending on the type of ascent, steep or flat—either the necessary highest climbing velocity or the required horizontal orbiting velocity is attained. The sooner this happens, the shorter the time during which the vehicle must be supported by the propulsion system and the lower the related propellant consumption will be. We see then that a high velocity must be attained as rapidly as possible during the ascent.

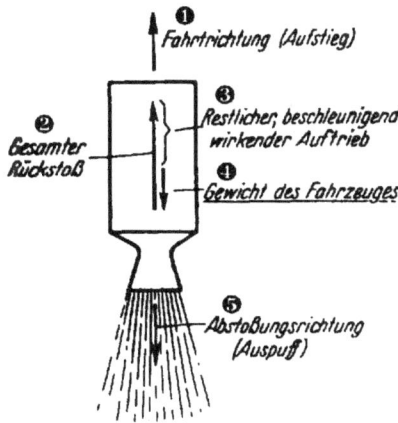

Figure 21. As long as the vehicle has to be supported (carried) by the propulsion system during the ascent, the forward thrust of the vehicle is decreased by its weight.

Key: 1. Direction of flight (ascent); 2. Total reactive force; 3. Remaining propulsive force available for acceleration; 4. Weight of the vehicle; 5. Direction of expulsion (exhaust).

However, a limit is soon set in this regard for space ships that are supposed to be suitable for transporting people. Because the related acceleration always results in the release of inertial forces during a forced velocity increase (as in this case for the propulsion system) and is not caused solely by the free interaction of the inertial forces. These forces are mani-

during a vertical ascent. In this case during the duration of propulsion, the vehicle and its contents would be subjected to the effect of the force of gravity of four times the strength of the Earth's normal gravity. Do not underestimate what this means! It means nothing less than that the feet would have to support almost four times the customary body weight.

Therefore, this ascent phase, lasting only a few minutes, can be spent by the passengers only in a prone position, for which purpose Oberth anticipated hammocks.

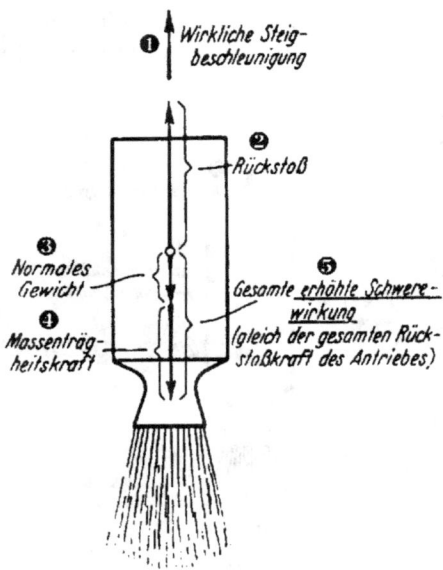

Figure 22. During the duration of propulsion, forces of inertia are activated in the vehicle due to the acceleration of the vehicle (increase in velocity) caused by propulsion; the forces manifest themselves for the vehicle like an increase in gravity.

Key: 1. Actual acceleration of climb; 2. Reaction; 3. Normal weight; 4. Force of inertia; 5. Total increased effect of gravity (equals the total reactive force of the propulsion system).

Taking into account the limitations in the magnitude of the acceleration, the highest climbing velocity that would be required for the total separation from the Earth can be attained only at an altitude of approximately 1,600 km with space ships occupied by humans during a vertical ascent. The rate of climb is then around 10,000 meters per second and is attained after somewhat more than 5 minutes. The propulsion system must be active that long. In accordance with what was stated previously, the vehicle is supported (carried) by the propulsion system during this time, and furthermore the resistance of the Earth's atmosphere still has to be overcome. Both conditions cause, however, an increase of the energy consumption such that the entire energy expenditure necessary for the ascent up to the total separation from the Earth finally becomes just as large as if an ideal highest velocity of around 13,000 meters per second would have to be imparted in total to the vehicle. Now this velocity (not the actual maximum climbing velocity of 10,000 meters per second) is critical for the amount of the propellants required.

Somewhat more favorable is the case when the ascent does not take place vertically, but on an inclined trajectory; in particular, when during the ascent the vehicle in addition strives to attain free orbital motion around the Earth as close to its surface as practical, taking the air drag into account (perhaps at an altitude of 60 to 100 km above sea level). And only then—through a further increase of the orbiting velocity—the vehicle works its way up to the highest velocity necessary for attaining the desired altitude or for the total separation from the Earth ("flat ascent," Figure 20).

The inclined direction of ascent has the advantage that the Earth's gravity does not work at full strength against the propulsion system (Figure 23), resulting, therefore, in a greater actual acceleration in the case of a uniform ideal acceleration (uniform propulsion)—which, according to what has been previously stated, is restricted when taking the well-being of the passengers into account. The greater acceleration results in the highest velocity necessary for the ascent being attained earlier.

However, the transition into the free orbital motion as soon as possible causes the vehicle to escape the Earth's gravity more rapidly than otherwise (because of the larger effect of the centrifugal force). Both conditions now cause the duration to be shortened during which the vehicle must be carried by the propulsion system, saving on the expenditure of energy as a result. Consequently, the ideal highest velocity to be imparted to the vehicle for totally separating from the Earth is only around 12,000 meters per second when employing this ascent maneuver, according to Oberth. In my opinion, however, we should come closest to the actually attainable velocity in practice when assuming an ideal highest velocity of approximately 12,500 meters per second.

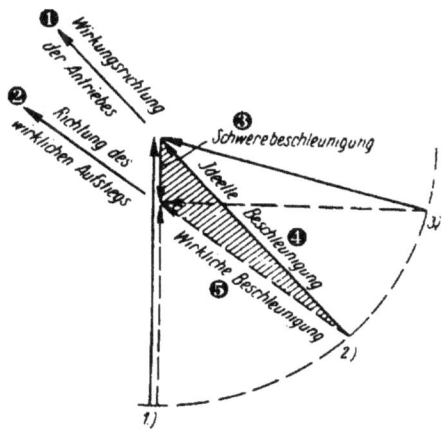

Figure 23. Acceleration polygon for: 1.) vertical ascent, 2.) inclined ascent, 3.) flat ascent. It can clearly be seen that the actual acceleration from 1.) to 3.) becomes greater and greater, despite a constant ideal acceleration (force of the propulsion system). (The acceleration polygon for 2.) is emphasized by hatched lines.)

Key: 1. Direction of the effect of the propulsion system; 2. Direction of the actual ascent; 3. Acceleration of gravity; 4. Ideal acceleration; 5. Actual acceleration.

Regardless of how the ascent proceeds, it requires in every case very significant accelerations, such that the vehicle attains a velocity of a projectile at an altitude of several kilometers. This condition—because of the thick density of the deepest layers of air closest to the surface of the

Earth—results in the air drag reaching undesirably high values in the very initial phases of the ascent, something that is particularly true for space rockets without people on board. Considerably greater accelerations of climb can be employed in unmanned vehicles than in manned ones because health is not a consideration for the former.

To come to grips with this disadvantage, the launch will take place from a point on the Earth's surface as high as possible, e.g., from a launch balloon or another air vehicle or from a correspondingly high mountain. For very large space ships, however, only the latter option is possible due to their weight, even though in this case the launch would preferably be carried out at a normal altitude.

General Comments About the Structure of the Space Rocket

Corresponding to the variety of purposes and goals possible for space ship flights, the demands placed on the vehicle will also be very different from mission to mission. For space ships, it will, therefore, be necessary to make the structure of the vehicle compatible with the uniqueness of the respective trip to a far greater extent than for the vehicles used for transportation to date. Nevertheless, the important equipment as well as the factors critical for the structure will be common for all space ships.

The external form of a space vehicle will have to be similar to that of a projectile. The form of a projectile is best suited for overcoming air drag at the high velocities attained by the vehicle within the Earth's atmosphere (projectile velocity, in accordance with previous statements!).

Fundamental for the internal structure of a rocket vehicle is the type of the propellants used. They must meet with the following requirements:

1. That they achieve an exhaust velocity as high as possible because the necessity was recognized previously for an expulsion velocity of the exhaust masses as high as possible.

2. That they have a density as high as possible (high specific weight), so that a small tank would suffice for storing the necessary amount of weight. Then, on the one hand, the weight of the tank is decreased and, on the other hand, the losses due to air drag also become smaller.

3. That their combustion be carried out in a safe way compatible with generating a constant forward thrust.

4. That handling them cause as few difficulties as possible.

Any type of gunpowder or a similar material (a solid propellant), such as used in fireworks rockets, would be the most obvious to use. The structure of the vehicle could then be relatively simple, similar to that of the familiar fireworks rocket. In this manner it would, no doubt, be possible to build equipment for various special tasks, and this would in particular pave the way for military technology, a point to be discussed below.

However for purposes of traveling in outer space, especially when the transportation of people is also to be made possible, using liquid propellants should offer far more prospects for development options, despite the fact that considerable engineering problems are associated with these types of propellants; this point will be discussed later.

The most important components of a space ship for liquid propellants are as follows: the propulsion system, the tanks for the propellants, the cabin and the means of landing. The propulsion system is the engine of the space ship. The reactive force is produced in it by converting the on-board energy stored in the propellant into forward thrust. To achieve this, it is necessary to pipe the propellants into an enclosed space in order to burn them there and then to let them discharge (exhaust) towards the rear. Two basic possibilities exist for this:

1. The same combustion pressure continuously exists in the combustion chamber. For the propellants to be injected, they must, therefore, be forced into the combustion chamber by overcoming this pressure. We will designate engines of this type as "constant pressure rocket engines."

2. The combustion proceeds in such a fashion that the combustion chamber is continuously reloaded in a rapid sequence with propellants, repeatedly ignited (detonated) and allowed to exhaust completely every time. In this case, injecting the propellants can also take place without an overpressure. Engines of this type we will designate as "detonation (or explosion) rocket engines."

The main components of the constant pressure rocket engines are the following: the combustion chamber, also called the firing chamber, and the nozzle located downstream from the combustion chamber (Figure 24). These components can exist in varying quantities, depending on the requirements.

The operating characteristics are as follows: the propellants (fuel and oxidizer) are forced into the combustion chamber in a proper state by means of a sufficient overpressure and are burned there. During the combustion, their chemically bonded energy is converted into heat and—in accordance with the related temperature increase—also into a pressure of the combustion gases generated in this manner and enclosed in the combustion chamber. Under the effect of this pressure, the gases of combustion escape out through the nozzle and attain as a result that velocity previously designated as "exhaust velocity." The acceleration of the gas molecules associated with this gain of velocity results, however, in the occurrence of counteracting forces of inertia (counter pressure, similar to pushing away an object![5]), whose sum now produces the force of

---

5. See page 13, Figure 12.

"reaction" (Figure 24) that will push the vehicle forward in the same fashion as has already been discussed earlier[6]. The forward thrust is obtained via heat, pressure, acceleration and reaction from the energy chemically bonded in the fuel.

So that this process is constantly maintained, it must be ensured that continually fresh propellants are injected into the combustion chamber. To this end, it is, however, necessary, as has been stated previously, that the propellant be under a certain overpressure compared to the combustion chamber. If an overpressure is supposed to be available in the tanks, then they would also have to have an appropriate wall thickness, a

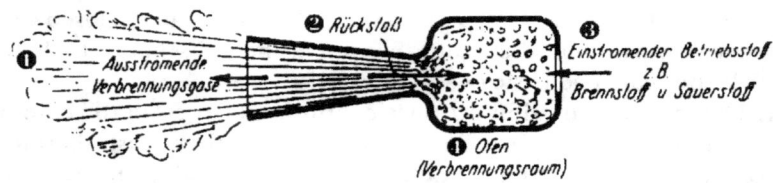

Figure 24. The combustion or firing chamber and the nozzle, the main components of the constant pressure rocket motor.

Key: 1. Escaping gases of combustion; 2. Reactive force; 3. Propellants flowing in, e.g., fuel and oxygen; 4. Combustion chamber.

property, however, that for larger tanks could present problems. Otherwise, pumps will have to be carried on board in order to put the propellants under the required pressure.

Furthermore, related equipment, such as injectors, evaporators and similar units are required so that the on-board liquid propellants can also be converted into the state suitable for combustion. Finally, the vehicle designers must also make provisions for sufficient cooling of the combustion chamber and nozzle, for control, etc.

The entire system has many similarities to a constant pressure gas turbine. And similar to that case, the not so simple question also exists in this case of a compatible material capable of withstanding high temperatures and of corresponding cooling options for the combustion chamber and nozzle. On the other hand, the very critical issue of a compressor for a gas turbine is not applicable for the rocket motor.

Similarly, the detonation rocket engine exhibits many similarities to the related type of turbine, the detonation (explosion) gas turbine. As with the latter, the advantage of a simpler propellant injection option must also be paid for in this case by a lower thermal efficiency and a

---

6. See page 15.

more complicated structure.

Which type of construction should be preferred can only be demonstrated in the future development of the rocket motor. Perhaps, this will also be, in part, a function of the particular special applications of the motor. It would not suffice to have only a motor functioning in completely empty space. We must still have the option of carrying on board into outer space the necessary amounts of energy in the form of propellant. Consequently, we are faced with a critically important question: the construction of the tanks for the fuel and oxidizer.

How large, in reality, is the amount of propellants carried on board? We know that the propulsion of the rocket vehicle occurs as a result of the fact that it continually expels towards the rear parts of its own mass (in our case, the propellants in a gasified state). After the propulsion system has functioned for a certain time, the initial mass of the vehicle (that is, its total mass in the launch-ready state) will have been decreased to a certain final mass by the amount of propellants consumed (and exhausted) during this time (Figure 25). This final mass represents, therefore, the total load that was transported by means of the amount of pro-

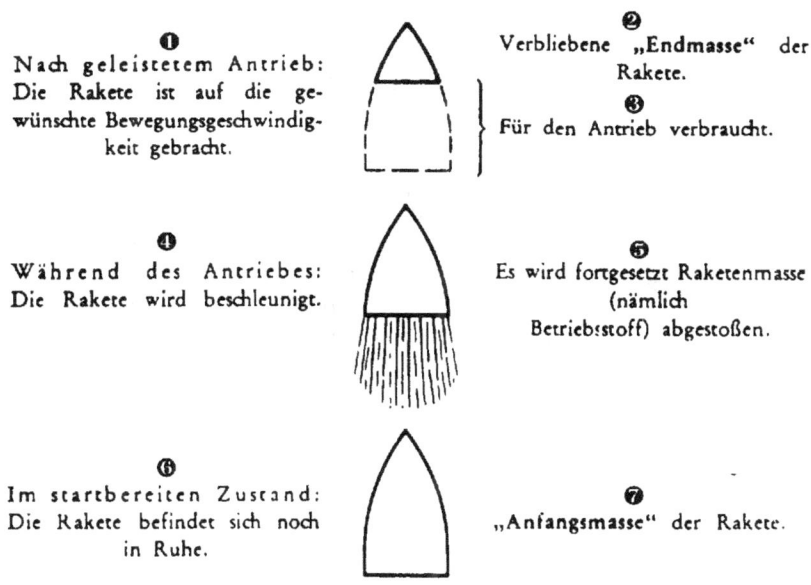

Figure 25.

Key: 1. Following a completed propulsion phase: The rocket is brought to the desired velocity of motion; 2. Remaining "final mass" of the rocket.; 3. Consumed for the propulsion; 4. During the propulsion phase: The rocket is accelerated; 5. Rocket mass (namely, the propellants) is continually expelled.; 6. In the launch-ready state: The rocket is at rest.; 7. "Initial mass" of the rocket.

pellants consumed, consisting of the payload, the vehicle itself and the remaining amounts of propellants.

The question is now as follows (Figure 26): How large must the initial mass M_0 be when a fixed final mass M is supposed to be accelerated to a velocity of motion v at a constant exhaust velocity c? The rocket equation provides an answer to this question: $M_0 = 2.72^{\frac{v}{c}} M$

According to the above, the initial mass M_0 of a space rocket is calculated as shown below. This mass should be capable of imparting the previously discussed[7] ideal highest climbing velocity of 12,500 meters per second, approximately necessary for attaining complete separation from the Earth.

M_0=520 M, for c=2,000 meters per second
M_0=64 M, for c=3,000 meters per second
M_0=23 M, for c=4,000 meters per second
M_0=12 M, for c=5,000 meters per second.

This implies the following: for the case that the exhaust velocity c is, by way of example, 3,000 meters per second, the vehicle, at the beginning of the propulsion phase, must be 64 times as heavy with the propellants necessary for the ascent as after the propellants are consumed. Consequently, the tanks must have a capacity to such an extent that they can hold an amount of propellants weighing 63 times as much as the empty space rocket, including the load to be transported, or expressed differently: an amount of propellants that is 98.5 percent of the total weight of the launch-ready vehicle.

Figure 26.

Key: 1. Velocity of motion; 2. Final mass; 3. Exhaust velocity; 4. Initial mass.

An amount of propellants of 22 times the weight would also suffice if the exhaust velocity is 4,000 meters per second and only 11 times if the exhaust velocity increases up to 5,000 meters per second. Ninety-six and 92 percent of the total weight of the launch-ready vehicle is allocated to the propellants in these two cases.

As has been frequently emphasized, the extreme importance of an expulsion (exhaust) velocity as high as possible can clearly be recognized from these values. (The velocity is the expression of the practical energy value of the propellant used!) However, only those rockets that are supposed to be capable of imparting the maximum climbing velocity

---

7. See pages 30-31.

necessary for the total separation from the Earth must have a propellant capacity as large as that computed above. On the other hand, the "ratio of masses" $\left(\text{ratio of the initial to the final mass of the rocket}: \dfrac{M_0}{M}\right)$ is considerably more favorable for various types of applications (explained later) in which lower highest velocities also suffice.

In the latter cases from a structural engineering point of view, fundamental difficulties would not be caused by the demands for the propellant capacity of the vehicle and/or of the tanks. By way of example, a space rocket that is supposed to attain the final velocity of v=4,200 meters per second at an exhaust velocity of c=3,000 meters per second would have to have a ratio of masses of $\dfrac{M_0}{M} = 4$, according to the rocket equation. That is, the rocket would have to be capable of storing an amount of propellant that is 75 percent of its total launch weight, a capability that can certainly be achieved from a structural engineering point of view.

To be sure, space rockets that can carry on board the amounts of propellants necessary for the complete separation from the Earth (according to what has already been stated, the amounts of propellants are 98.5 percent of the launch weight at an exhaust velocity of c=3,000 meters per second), could, for all practical purposes, not be easily realized. Fortunately, there is a trick making it possible to circumvent this structural difficulty in a very simple manner: the so-called staging principle that both Goddard and Oberth recognized independently of one another as a fundamental principle of rocket technology.

In accordance with this principle, the desired final velocity need not be attained with a single rocket; but rather, the space rocket is divided into multiple units (stages), each one always forming the load for the next largest unit. If, for example, a three-stage space rocket is used, then it consists of exactly three subrockets: the subrocket 3 is the smallest and carries the actual payload. It forms (including this payload) the load of subrocket 2 and the latter again (including subrocket 3 and its payload) the load of subrocket 1. During ascent, subrocket 1 functions first. As soon as this stage is used up, its empty shell is decoupled and subrocket 2 starts to function. When it is spent, it also remains behind and now subrocket 3 functions until the desired final velocity is attained. Only the latter arrives at the destination with the payload.

Because the final velocities of three subrockets are additive in this process, each individual one must be able to generate only $1/3$ of the total required final velocity. In the case of a 3-stage space rocket, which is supposed to attain the highest climbing velocity of 12,500 meters per second necessary for the total separation from the Earth, only a final velocity to be attained of around 4,200 meters per second would consequently be allocated to each subrocket. For that, however, the propellant capacity, certainly implementable from an engineering point of view, of 75 percent $\left(\text{ratio of the masses } \frac{M_0}{M}=4\right)$ suffices, as we determined previously, at an exhaust velocity of c=3,000 meters per second, for example. If the individual subrockets can, however, be manufactured, then no doubt exists about the possibility of erecting the complete rocket assembled from all subrockets.

As a precautionary measure, let's examine the absolute values of the rocket masses or rocket weights resulting from the above example. Assume a payload of 10 tons is to be separated from the Earth; the individual subrockets may be built in such a fashion that their empty weight is as large as the load to be transported by them. The weights of the subrockets in tons result then as follows:

The initial weight of the total space rocket consisting of 3 stages would be 5,120 tons, a number that is not particularly impressive, considering the fact that technology is capable of building, for example, an ocean liner weighing 50,000 tons.

In this fashion—by means of the staging principle—it would actually be possible to attain any arbitrary final velocity, in theory at least. For all practical purposes in this regard, fixed limitations will, of course, result, in particular when taking the absolute values of the initial weights into

Subrocket	Load	Empty weight	Final weight M	Initial weight M_0
3	10	10	10 + 10=20[1]	4 × 20=80[2]
2 + 3	80	80	80 + 80=160	4 × 160=640
1 + 2 + 3	640	640	640 + 640=1280	4 × 1280=**5120**

1) The final weight M is equal to the empty weight plus the load when the rocket—as in this case—functions until its propellants are completely consumed.

2) The initial weight M_0 is, in this case, equal to 4 times the final weight M because, as has been stated previously in our example, each subrocket approaches the ratio of masses (weights) $\frac{M_0}{M}=4$.

consideration. Nevertheless an irrefutable proof is inherent in the staging principle to the effect that it would be fundamentally possible to build space rockets capable of separating from the Earth even with the means available today.

That does not mean the staging principle represents the ideal solution for constructing space rockets in the described form, because it leads to an increase of the dead weight and, as a result, of the propellants necessary for transportation. This, however, is not now a critical point. Initially, we are only concerned with showing "that it is possible in the first place." Without a doubt every type of space rocket construction, regardless of which one, will have to employ the fundamental concept expressed in the staging principle: during the duration of propulsion—for the purpose of saving propellants—every part of the vehicle that has become unnecessary must be immediately released (jettisoned) in order not to carry dead weight uselessly and, at the same time, to have to accelerate further with the remaining weight. It is assumed, of course, that we are dealing with space rockets that are supposed to attain greater final velocities.

From a structural engineering point of view, we do not want to conceal the fact that certainly quite a few difficulties will arise as a result of the still significant demands imposed on the capacity of the propellant tanks—despite the staging principle. In this regard, it will be necessary in part to use construction methods deviating fundamentally from the customary ones, because all parts of the vehicle, in particular the tanks, must be made as light-weight as possible. Nevertheless, the tanks must have sufficient strength and stiffness to be able to withstand both the pressure of mass and the atmospheric stagnation pressure during the ascent, taking into account that many of the usual metals become brittle and, therefore, lose strength at the extreme lower temperatures to which the tanks may be exposed.

Moreover in a space ship, a compartment (cell) must exist for housing the pilot and passengers and for storing supplies of the life support necessities and other equipment, as well as for storing freight, scientific devices for observations, etc. The compartment must be air-sealed and must have corresponding precautionary measures for artificially supplying air for breathing and for maintaining a bearable temperature. All equipment necessary for controlling the vehicle is also stored in the compartment, such as manual controls for regulating the propulsion system; recorders for time, acceleration, velocity, and path (altitude); and for determining the location, maintaining the desired direction of flight, and similar functions. Even space suits (see the following), hammocks, etc. must be available. Finally, the very important aids for landing, such as parachutes, wings, etc. also belong to the equipment of a space ship.

Proposals To Date

The following are the various recommendations made to date for the practical solution of the space flight problem:

Professor Goddard uses a smokeless powder, a solid substance, as a propellant for his space rockets. He has not described any particular device, but recommends only in general packing the powder into cartridges and injecting it automatically into the combustion chamber, in a fashion similar to that of a machine gun. The entire rocket should be composed of individual subrockets that are jettisoned one after the other during the ascent, as soon as they are spent, with the exception of that subrocket containing the payload, and it alone reaches the destination. First of all, he intends to make unmanned devices climb to an altitude of several hundred kilometers. Subsequently, he also wants to try to send up an unmanned rocket to the Moon carrying only several kilograms of luminous powder. When landing on the Moon, the light flare is supposed to flash, so that it could then be detected with our large telescopes, thus verifying the success of the experiment. Reportedly, the American Navy is greatly interested in Goddard's devices.

The results of practical preliminary experiments conducted and published by Goddard to date are very valuable; the means for carrying out these experiments were provided to him in a very generous manner by the famous Smithsonian Institution in Washington. He was able to attain exhaust velocities up to 2,434 meters per second with certain types of smokeless powder when appropriately shaping and designing the nozzles. During these experiments, he was successful in using 64.5 percent of the energy chemically bonded in the powder, that is, to convert it into kinetic energy of the escaping gases of combustion. The result agrees approximately with the experiences of ballistics, according to which about $2/3$ of the energy content of the powder can be used, while the remainder is carried as heat by the exhaust gases and, as a result, is lost. Perhaps, the efficiency of the combustion chamber and nozzle can be increased somewhat during further engineering improvements, to approximately 70 percent.

Therefore, an "internal efficiency" of approximately 60 percent could be expected for the entire propulsion system—the rocket motor—after taking into consideration the additional losses caused by the various auxiliary equipment (such as pumps and similar devices) as well as by other conditions. This is a very favorable result considering that the efficiency is hardly more than 38 percent even for the best thermal engines known to date.

It is a good idea to distinguish the internal efficiency just considered from that addressed previously: the efficiency of the reactive force,[8] which could also be designated as the "external efficiency" of the rocket motor to distinguish it from the internal efficiency. Both are completely independent from one another and must be considered at the same time in order to obtain the total efficiency of the vehicle (which is just the product of the internal and external efficiency). As an example, the values of the efficiency for benzene as the fuel are listed in Column 3 of Table 2, page 22.

*

Differing from Goddard, Professor Oberth suggests using liquid propellants, primarily liquid hydrogen and also alcohol, both with the amounts of liquid oxygen necessary for their combustion. The hydrogen-oxygen mixture—called "detonating gas"—has the highest energy content (3,780 calories per kilogram compared to approximately 1,240 for the best smokeless powder) per unit of weight of all known substances. Accordingly, it yields by far the highest exhaust velocity. Oberth figured being able to attain approximately 3,800-4,200 meters per second. If we were successful in using the energy chemically bonded in detonating gas up to the theoretically highest possible limit, then its exhaust velocity could even exceed 5,000 meters per second. The gas resulting from the combustion is water vapor.

Unfortunately, the difficulty of carrying and using the gas in a practical sense is a big disadvantage compared to the advantage of its significant energy content and therefore relatively high exhaust velocity, due to which the detonating gas would in theory appear to be by far the most suitable propellant for space rockets. Storing hydrogen as well as oxygen in the rocket is possible only in the liquefied state for reasons of volume.

However, the temperature of liquid oxygen is -183°, and that of the liquid hydrogen only -253° Celsius. It is obvious that this condition must considerably complicate the handling, even disregarding the unusual requirements being imposed on the material of the tanks. Additionally, the average density (specific weight) of detonating gas is very low even in a liquefied state so that relatively large tanks are necessary for storing a given amount of the weight of the gas.

In the case of alcohol, the other fuel recommended by Oberth, these adverse conditions are partially eliminated but cannot be completely avoided. In this case, the oxygen necessary for combustion must also be carried on board in the liquid state. According to Oberth, the exhaust velocity is approximately 1,530-1,700 meters per second for alcohol, con-

8. See pages 18-19.

siderably lower than for hydrogen. It does have a greater density, however.

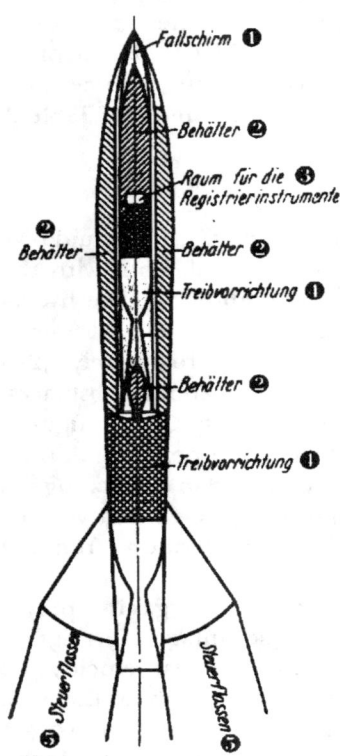

Figure 27. A longitudinal cross section through the main rocket of Oberth's small rocket model is shown schematically. The hydrogen rocket is inserted in the forward part of the alcohol rocket.

Key: 1. Parachute; 2. Tank; 3. Space for the recording instruments; 4. Propulsion system; 5. Control fins.

Due to these properties, Oberth uses alcohol together with liquid oxygen as propellants for the initial phase of the ascent, because the resistance of the dense layers of air near the Earth's surface must be overcome during the ascent. Oberth viewed a large cross-sectional loading (i.e., the ratio of the total mass of a projectile to the air drag cross section of the projectile) as advantageous even for rockets and recommended, besides other points: "to increase the mass ratio at the expense of the exhaust velocity."[9] This is, however, attained when alcohol and oxygen are used as propellants.

Oberth's space rocket has, in general, the external shape of a German S-projectile and is composed of individual subrockets that are powered either with hydrogen and oxygen (hydrogen rocket) or with alcohol and oxygen (alcohol rocket). Oberth also described in more detail two examples of his space vehicle. Of the two, one is a smaller, unmanned model, but equipped with the appropriate recording instruments and is supposed to ascend and perform research on the higher and highest layers of air. The other one is a large space ship designed for transporting people.

The smaller model (Figure 27) consists of a hydrogen rocket that is inserted into the forward part of a considerably larger alcohol rocket. Space for storing the re-

---

9. However, we can not support this suggestion, as must be particularly emphasized in the present case. The suggestion can hardly be tenable because it is based on the assumption that the concept of the "cross sectional loading" used in ballistics could also be applied in this case. However, in our opinion, the latter is not really acceptable; the rocket moving forward with propulsion is subject to mechanical conditions that are substantially different from those of a ballistic projectile.

cording instruments is located below the tank of the hydrogen rocket. At the end of the alcohol rocket, movable fins are arranged that are supposed to stabilize and to control the vehicle. The entire apparatus is 5 meters long, measures 56 cm in diameter and weighs 544 kg in the launch-ready state.

Furthermore, a so-called "booster rocket" (Figure 28) is provided that is 2 meters high, 1 meter in diameter and weights 220 kg in the launch-ready state. Launching takes place from dirigibles at an altitude of 5,500 meters or more (Figure 29). Initially the booster rocket, which later will be jettisoned, lifts the main rocket to an altitude of 7,700 meters and accelerates it to a velocity of 500 meters per second (Figure 30). Now, the rocket is activated automatically: first the alcohol rocket and, after it is spent and decoupled, the hydrogen rocket. Fifty-six seconds after the launch, a highest climbing velocity of 5,140 meters per second is attained, which suffices for the remaining hydrogen rocket, now without propulsion, to reach a final altitude of approximately 2,000 km in a free ascent. The return to Earth takes place by means of a self-deploying parachute stored in the tip of the hydrogen rocket.

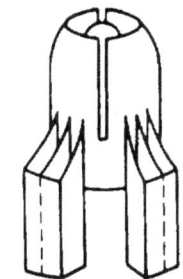

Figure 28. The booster rocket of Oberth's small rocket model.

Figure 29. Launching the rocket from dirigibles, according to Oberth.

In the case of the second model, the large rocket space ship designed for transporting people (Figure 31), the total forward part of the vehicle consists of a hydrogen rocket set atop an alcohol rocket in the rear. The cabin designed for passengers, freight, etc. and equipped with all control devices, is located in the forward part of the hydrogen rocket. The parachute is stored above it. Towards the front, the vehicle has a metal cap shaped like a projectile, which later is jettisoned as unnecessary ballast along with the alcohol rocket (Figure 32), because the Earth's atmosphere is left behind at this point, i.e., no further air drag must be overcome. From here on, stabilization and controlling is no longer achieved by means of fins, but by control nozzles.

For this model, launching is performed over the ocean. In this case, the alcohol rocket operates first. It accelerates the vehicle to a climbing velocity of 3,000-4,000 meters per second, whereupon it is decoupled and left behind (Figure 32); the hydrogen rocket then begins to work in order to impart to the vehicle the necessary maximum climbing velocity or, if necessary, also a horizontal orbital velocity. A space ship of this nature, designed for transporting an observer, would, according to Oberth's data, weigh 300 metric tons in the launch-ready state.

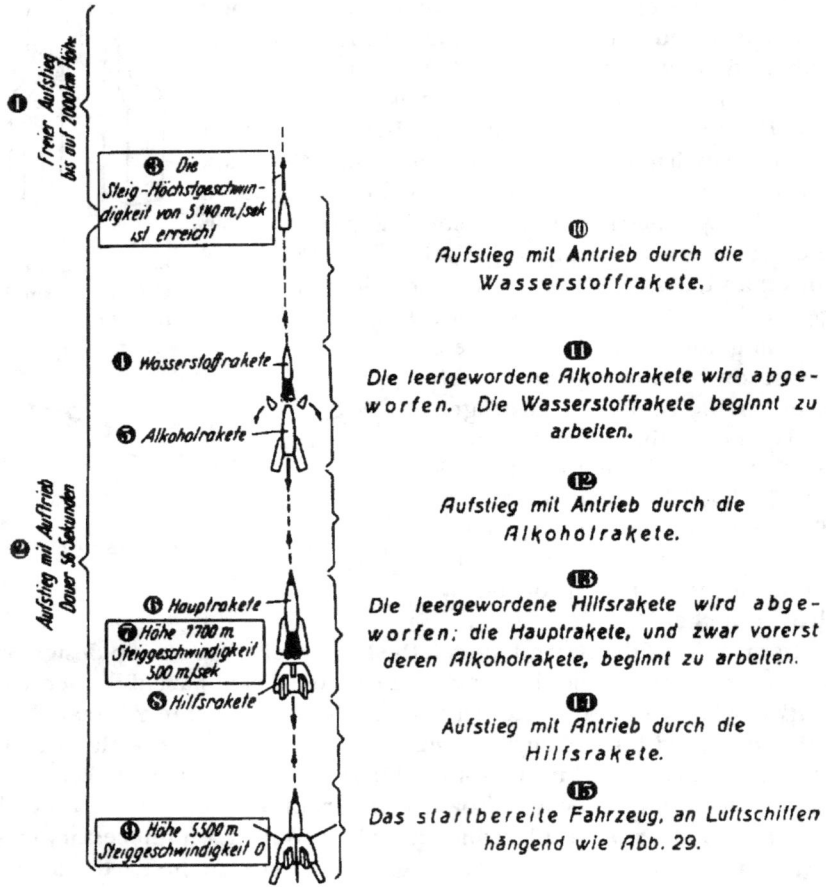

Figure 30. The ascent of Oberth's small (unmanned) rocket model.

Key: 1. Free ascent up to an altitude of 2,000 km; 2. Powered ascent lasting 56 seconds; 3. The highest climbing velocity of 5,140 m/sec is attained; 4. Hydrogen rocket; 5. Alcohol rocket; 6. Complete rocket; 7. Altitude of 7,700 m, climbing velocity of 500 m/sec; 8. Booster rocket; 9. Altitude of 5,500 m, climbing velocity of 0; 10. Powered ascent by the hydrogen rocket; 11. The empty alcohol rocket is jettisoned. The hydrogen rocket starts to operate; 12. Power ascent by the alcohol rocket; 13. The empty booster rocket is jettisoned; the main rocket, beginning with its alcohol rocket, starts to operate; 14. Powered ascent by the booster rocket; 15. The launch-ready vehicle, suspended from dirigibles, as shown in Figure 29.

In both models, the subrockets are independent; each has, therefore, its own propulsion system as well as its own tanks. To save weight, the latter are very thin-walled and obtain the necessary stiffness through inflation, that is, by the existence of an internal overpressure, similar to non-rigid dirigibles. When the contents are being drained, this overpressure is maintained by evaporating the remaining liquid. The oxygen tank is made of copper and the hydrogen tank of lead, both soft metals, in order to prevent the danger of embrittlement caused by the extreme low temperatures discussed previously.

The propulsion equipment is located in the rear part of each rocket (Figure 33). For the most part, that equipment consists of the combustion chamber and one or more thin sheet metal exhaust nozzles connected to it, as well as various pieces of auxiliary equipment necessary for propulsion, such as injectors and other devices. Oberth uses unique pumps of his own design to produce the propellant overpressure necessary for injection into the combustion chamber. Shortly before combustion, the oxygen is gasified, heated to 700° and then blown into the chamber, while the fuel is sprayed into the hot oxygen stream in a finely dispersed state. Arrangements are made for appropriately cooling the chamber, nozzles, etc.

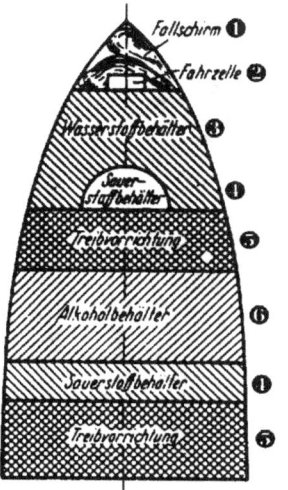

Figure 31. A longitudinal cross section of Oberth's large rocket for transporting people is shown schematically. The hydrogen rocket is set atop the alcohol rocket.

Key: 1. Parachute; 2. Cabin; 3. Hydrogen tank; 4. Oxygen tank; 5. Propulsion system; 6. Alcohol tank.

It should be noted how small the compartment for the payload is in comparison to the entire vehicle, which consists principally of the tanks. This becomes understandable, however, considering the fact that the amounts of propellants previously calculated with the rocket equation[10] and necessary for the ascent constitute as much as 20 to 80 percent of the total weight of the vehicle, propellant residuals, and payload!

However, the cause for this enormous propellant requirement lies not in an insufficient use of the propellants, caused perhaps by the deficiency of the reaction principle used for the ascent, as is frequently and incorrectly thought to be the case. Naturally, energy is lost during the

---

10. See pages 36-39.

ascent, as has previously been established,[11] due to the circumstance that the travel velocity during the propulsion phase increases only gradually and, therefore, is not of an equal magnitude (namely, in the beginning smaller, later larger) with the exhaust (repulsion) velocity (Figure 17). Nevertheless, the average efficiency of the reaction[12] would be 27 percent at a constant exhaust velocity of 3,000 meters per second and 45 percent at a constant exhaust velocity of 4,000 meters per second, if, for example, the vehicle is supposed to be accelerated to the velocity of 12,500 meters per second, ideally necessary for complete separation from the

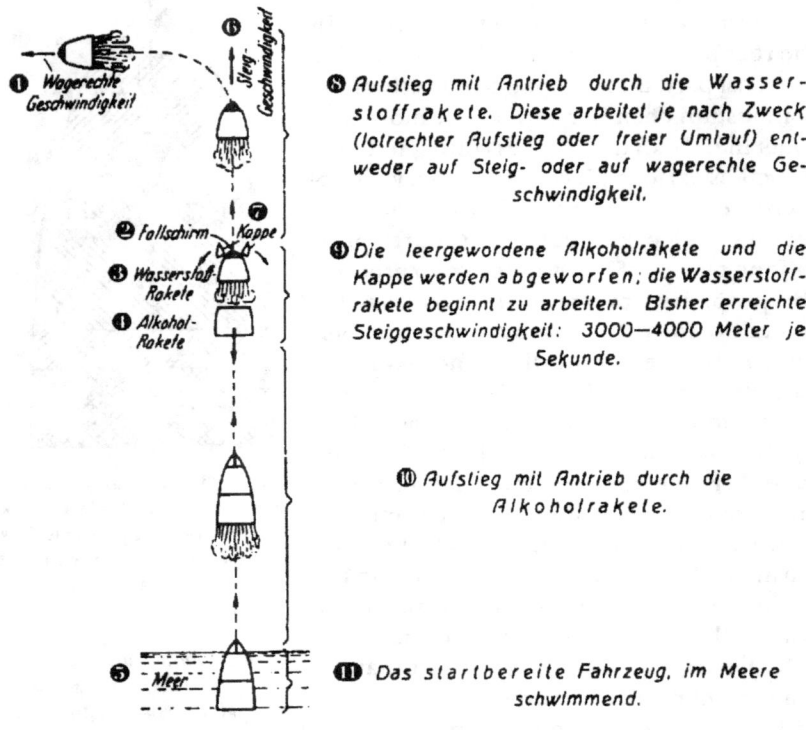

Figure 32. The ascent of Oberth's larger (manned) rocket model.

Key: 1. Horizontal velocity; 2. Parachute; 3. Hydrogen rocket; 4. Alcohol rocket; 5. Ocean; 6. Climbing velocity; 7. Cap; 8. Powered ascent by the hydrogen rocket. Depending on the purpose (vertical ascent or free orbiting), this rocket imparts either a climbing velocity or a horizontal velocity; 9. The empty alcohol rocket and the cap are jettisoned; the hydrogen rocket starts to operate. The climbing velocity attained up to this point is 3,000 to 4,000 meters per second; 10. Powered ascent by the alcohol rocket; 11. The launch-ready vehicle floating in the ocean.

11. See pages 25-29.
12. Using the formula on pages 27-28.

Earth. According to our previous considerations, the efficiency would even attain a value of 65 percent in the best case, i.e., for a propulsion phase in which the final velocity imparted to the vehicle is 1.59 times the exhaust velocity.[13]

Since the internal efficiency of the propulsion equipment can be estimated at approximately 60 percent on the basis of the previously discussed Goddard experiments and on the experiences of ballistics[14], it follows that an average total efficiency of the vehicle of approximately 16 to 27 percent (even to 39 percent in the best case) may be expected during the ascent, a value that, in fact, is no worse than for our present day automobiles! Only the enormous work necessary for overcoming such vast altitudes requires such huge amounts of propellants.

If, by way of example, a road would lead from the Earth into outer space up to the practical gravitational boundary, and if an automobile were supposed to drive up that road, then an approximately equal supply of propellants, including the oxygen necessary for combustion, would have to be carried on the automobile, as would be necessary for the propellants of a space ship with the same payload and altitude.

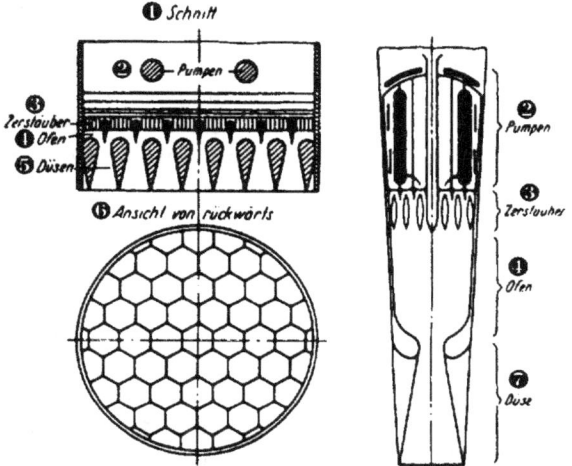

Figure 33. The propulsion system of Oberth's rocket: Right: the small model. The combustion chamber discharges into only one nozzle. Left: the large model. A common combustion chamber discharges into many nozzles arranged in a honeycombed fashion.

Key: 1. Sectional view; 2. Pumps; 3. Injectors; 4. Combustion chamber; 5. Nozzles; 6. View from the rear; 7. Nozzle.

It is also of interest to see how Oberth evaluated the question of costs. According to his data, the previously described smaller model including the preliminary experiments would cost 10,000 to 20,000 marks. The

13. See Table 4 on page 28.
14. See page 40.

construction costs of a space ship, suitable for transporting one observer, would be over 1 million marks. Under favorable conditions, the space ship would be capable of carrying out approximately 100 flights. In the case of a larger space ship, which transports, besides the pilot together with the equipment, 2 tons of payload, an ascent to the stable state of suspension (transition into a free orbit) would require approximately 50,000 to 60,000 marks.

*

The study published by Hohmann about the problem of space flight does not address the construction of space rockets in more detail, but rather thoroughly addresses all fundamental questions of space flight and provides very valuable recommendations related to this subject. As far as questions relating to the landing process and distant travel through outer space are concerned, they will be addressed later.

What is interesting at this point is designing a space vehicle for transporting two people including all necessary equipment and supplies. Hohmann conceives a vehicle structured in broad outlines as follows: the actual vehicle should consist only of the cabin. In the latter, everything is stored—with the exception of the propellant. A solid, explosive-like substance serving as the propellant would be arranged below the cabin in the shape of a spire tapering upward in such a way that the cabin forms its peak (Figure 34). As a result of a gradual burning of this propellant spire, thrust will be generated similar to that of a fireworks rocket. A prerequisite for this is that explosive experts find a substance that, on the one hand, has sufficient strength to keep itself in the desired shape and that, on the other hand, also has the energy of combustion necessary for generating a relatively large exhaust velocity.

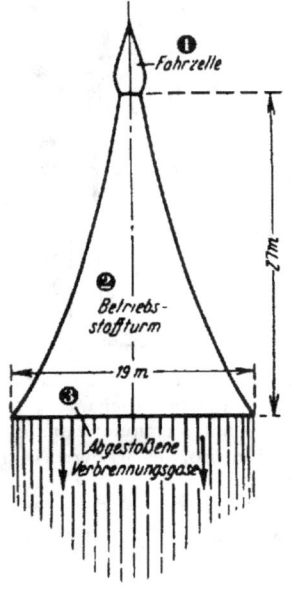

Figure 34. The space rocket according to Hohmann.

Key: 1. Cabin cell; 2. Propellant tower; 3. Exhaust gases of combustion.

Assuming that this velocity is 2,000 meters per second, a space vehicle of this nature would weigh, according to Hohmann, a total of 2,800 tons in the launch-ready state, if it is to be capable of attaining an altitude of 800,000 km (i.e., twice the distance to the Moon). This corresponds approximately to the weight of a small ocean liner. A round trip of this nature would last 30.5 days.

Recent publications by von Hoefft are especially noteworthy. His original thought was to activate the propulsion system of space ships using the space ether. For this purpose, a uni-directional ether flow is supposed to be forced through the vehicle by means of an electrical field. Under Hoefft's assumption, the reaction effect of the ether would then supply the propulsive force of the vehicle, a concept that assumes ether has mass. Hoefft, however, maintained that was assured if the opinion held by Nernst and other researchers proved to be correct. According to this view, the space ether should possess a very significant internal energy (zero point energy of the ether); this was believed to be substantiated by the fact that energy is also associated with mass in accordance with Einstein's Law.*

Taking into account the improbability of being able to implement these concepts in the foreseeable future, von Hoefft has now agreed with Oberth's efforts. According to reports, his latest research on this subject has resulted in developing designs that are only awaiting funding.

He intends initially to launch an unmanned recording rocket to an altitude of approximately 100 km for the purpose of exploring the upper layers of the atmosphere. This rocket has one stage, is powered by alcohol and liquid oxygen, and is controlled by means of a gyroscope like a torpedo. The height of the rocket is 1.2 meters, its diameter is 20 cm, its initial (launch) weight is 30 kg and its final weight is 8 kg, of which 7 kg are allocated to empty weight and 1 kg to the payload. The latter is composed of a meteorograph stored in the top of the rocket and separated automatically from the rocket as soon as the final altitude is attained, similar to what happens in recording balloons. The meteorograph then falls alone slowly to Earth on a self-opening parachute, recording the pressure, temperature and humidity of the air. The ascent is supposed to take place at an altitude of 10,000 meters from an unmanned rubber balloon (pilot balloon) to keep the rocket from having to penetrate the lower, dense layers of air.

As the next step, von Hoefft plans to build a larger rocket with an initial weight of 3,000 kg and a final weight of 450 kg, of which approxi-

---

* Walther Hermann Nernst (1864-1941) was one of the founders of modern physical chemistry. He won the Nobel Prize for chemistry in 1920 for formulating the third law of thermodynamics. Albert Einstein (1879-1955) was, of course, the physicist who formulated the special and general theories of relativity, the photon theory of light, and the equivalence of mass and energy (expressed in the famous formula, $e=mc^2$). He won the Nobel Prize in physics in 1921. Potocnik seems not to be aware that Einstein's special theory of relativity made the ether hypothesis—that ether was a universal substance acting as a medium for the transmission of electromagnetic waves—obsolete.

mately 370 kg are allocated to empty weight and 80 kg to the payload. Similar to a projectile, the rocket is supposed to cover vast distances of the Earth's surface (starting at approximately 1,500 km) in the shortest time on a ballistic trajectory (Keplerian ellipses) and either transport mail or similar articles or photograph the regions flown over (for example, the unexplored territories) with automatic camera equipment.

Landing is envisaged in such a manner that the payload is separated automatically from the top before the descent, similar to the previously described recording rocket, descending by itself on a parachute.

This single-stage rocket could also be built as a two-stage rocket and as a result be made appropriate for a Moon mission. For this purpose, it is equipped, in place of the previous payload of approximately 80 kg, with a second rocket of the same weight; this rocket will now carry the actual, considerably smaller payload of approximately 5 to 10 kg. Because the final velocities of both subrockets in a two-stage rocket of this type are additive in accordance with the previously explained staging principle,[15] a maximum climbing velocity would be attained that is sufficiently large to take the payload, consisting of a load of flash powder, to the Moon. When landing on the Moon, this load is supposed to ignite, thus demonstrating the success of the experiment by a light signal, as also proposed by Goddard. Both this and the aforementioned mail rocket are launched at an altitude of 6,000 meters from a pilot balloon, a booster rocket, or a mountain top.

In contrast to these unmanned rockets, the large space vehicles designed for transporting people, which Hoefft then plans to build in a follow-on effort, are supposed to be launched principally from a suitable body of water, like a seaplane, and at the descent, land on water, similar to a plane of that type. The rockets will be given a special external shape (somewhat similar to a kite) in order to make them suitable for their maneuvers.

The first model of a space vehicle of this type would have a launch weight of 30 tons and a final weight of 3 tons. Its purpose is the following: on the one hand, to be employed similarly to the mail rocket yet occupied by people who are to be transported and to cover great distances of the Earth's surface on ballistic trajectories (Keplerian ellipses) in the shortest time; and, on the other hand, it would later have to serve as an upper stage of larger, multi-stage space ships designed for reaching distant celestial bodies. Their launch weights would be fairly significant: several hundred metric tons, and even up to 12,000 tons for the largest designs.

---

15. See pages 37-39.

Comments Regarding Previous Design Proposals

Regarding these various proposals, the following is added as supplementary information: as far as can be seen from today's perspective, the near future belongs in all probability to the space rocket with liquid propellants. Fully developed designs of such rockets will be achieved when the necessary technical conditions have been created through practical solutions (obtained in experiments) of the questions fundamental to their design: 1. methods of carrying the propellants on board, 2. methods of injecting propellants into the combustion chamber, and 3. protection of the chamber and nozzle from the heat of combustion.

For this reason, we intentionally avoided outlining our own design recommendations here. Without a doubt, we consider it advisable and necessary, even timely, at least as far as it is possible using currently available experiences, to clarify the fundamentals of the vehicle's structure; the question of propellant is predominantly in this context. As stated earlier, hydrogen and oxygen, on the one hand, and alcohol and oxygen, on the other, are suggested as propellants.

In the opinion of the author, the pure hydrocarbon compounds (together with the oxygen necessary for combustion) should be better suited than the ones mentioned in the previous paragraph as propellants for space rockets. This becomes understandable when the energy content is expressed as related to the volume instead of to the weight, the author maintaining this as being the most advantageous method in order to be able to evaluate the value of a rocket fuel in a simple fashion. Not only does it matter what amount of fuel by weight is necessary for a specific performance; still more important for storing the fuel, and as a result for designing the vehicle, is what amount of fuel by volume must be carried on board. Therefore, the energy content (thermal units per liter) of the fuel related to the volume provides the clearest information.

This energy content is the more significant the greater the specific weight as well as the net calorific value of the fuel under consideration are, and the less oxygen it requires for its combustion. In general, the carbon-rich compounds are shown to be superior to the hydrogen-rich ones, even though the calorific value per kilogram of the latter is higher. Consequently, benzene would appear very suitable, for example. Pure carbon would be the best. Because the latter, however, is not found in the fluid state, attempts should be made to ascertain whether by mechanical mixing of a liquid hydrocarbon (perhaps benzene, heptane, among others) with an energy content per liter as high as possible with finely-dispersed carbon as pure as possible (for instance carbon black, the finest coal dust or similar products), the energy content per liter could be increased still further and as a result particularly high quality rocket fuel could be obtained, which may perhaps be overall the best possible in accordance with our current knowledge of substances.

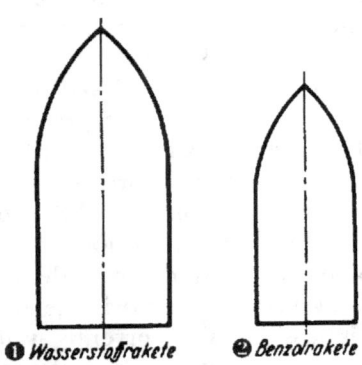

Figure 35. Size relationship between a hydrogen rocket and a benzene rocket of the same performance, when each one is supposed to be capable of attaining a velocity of 4,000 meters per second.

Key: 1. Hydrogen rocket; 2. Benzene rocket.

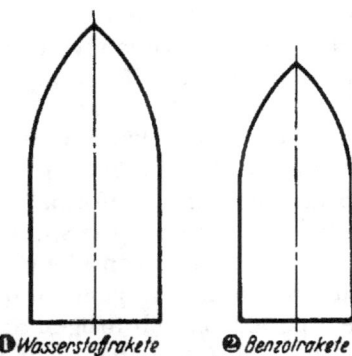

Figure 36. Size relationship between a hydrogen rocket and a benzene rocket of the same performance, when each one is supposed to be capable of attaining a velocity of 12,500 meters per second (complete separation from the Earth!).

Key: 1. Hydrogen rocket; 2. Benzene rocket

Of course, an obvious condition for the validity of the above considerations is that all fuels work with the same efficiency. Under this assumption by way of example, a space rocket that is supposed to attain the final velocity of 4,000 meters per second would turn out to be smaller by about one-half and have a tank surface area smaller by one-third when it is powered with benzene and liquid oxygen than when powered by liquid hydrogen and oxygen (Figure 35).

Therefore, the benzene rocket would not only be realized sooner from an engineering point of view, but also constructed more cheaply than the hydrogen rocket of the same efficiency, even though the weight of the necessary amount of fuel is somewhat higher in the former case and, therefore, a larger propulsion force and, consequently, stronger, heavier propulsion equipment would be required. Instead, the fuel tanks are smaller for benzene rockets and, furthermore, as far as they serve the purposes of benzene at least, can be manufactured from any lightweight metal because benzene is normally liquid. When considering its abnormally low temperature (-253° Celsius) according to Oberth, a point made previously, rockets for liquid hydrogen would have to be made of lead (!). This discussion ignores completely the many other difficulties caused by this low temperature in handling liquid hydrogen and the method of using this fuel; all of these difficulties disappear when using benzene.

However, this superiority of liquid

hydrocarbons compared to pure hydrogen diminishes more and more at higher final velocities. Nevertheless, a benzene rocket would still turn out to be smaller by one-third than a hydrogen rocket, even for attaining a velocity of 12,500 meters per second—as is ideally necessary for complete separation from the Earth (Figure 36). Only for the final velocity of 22,000 meters per second would the volumes of propellants for the benzene rocket be as large as for the hydrogen rockets. Besides these energy-efficient advantages and other ones, liquid hydrocarbons are also considerably cheaper than pure liquid hydrogen.

The Return to Earth

The previous explanations indicate that obstacles stand in the way of the ascent into outer space which, although significant, are nonetheless not insurmountable. Based solely on this conclusion and before we address any further considerations, the following question is of interest: whether and how it would be possible to return to Earth after a successful ascent and to land there without experiencing any injuries. It would arouse a terrible horror even in the most daring astronaut if he imagined, seeing the Earth as a distant sphere ahead of him, that he will land on it with a velocity of no less than approximately 12 times the velocity of an artillery projectile as soon as he, under the action of gravity, travels towards it or more correctly stated, crashes onto it.

The rocket designer must provide for proper braking. What difficult problem is intrinsic in this requirement is realized when we visualize that a kinetic energy, which about equals that of an entire express train moving at a velocity of 70 km/hour, is carried by each single kilogram of the space ship arriving on Earth! For, as described in the beginning, an object always falls onto the Earth with the velocity of approximately 11,000 meters per second when it is pulled from outer space towards the Earth by the Earth's gravitational force. The object has then a kinetic energy of around 6,000 metric ton-meters per kilogram of its weight. This enormous amount of energy must be removed in its entirety from the vehicle during braking.

Only two possibilities are considered in this regard: either counteracting the force by means of reaction propulsion (similar to the "reverse force" of the machine when stopping a ship), or braking by using the Earth's atmosphere. When landing according to the first method, the propulsion system would have to be used again, but in an opposite direction to that of flight (Figure 37). In this regard, the vehicle's descent energy would be removed from it by virtue of the fact that this energy is offset by the application of an equally large, opposite energy. This requires, however, that the same energy for braking and, therefore, the same amount of fuel necessary for the ascent would have to be consumed. Then, since the initial velocity for the ascent (highest climbing velocity) and the final velocity during the return (descent velocity) are of similar magnitudes, the kinetic energies, which must be imparted to the vehicle in the former case and removed in the latter case, differ only slightly from one another.

For the time being, this entire amount of fuel necessary for braking must still—and this is critical—be lifted to the final altitude, something that means an enormous increase of the climbing load. As a result, however, the amount of fuel required in total for the ascent becomes now so large that this type of braking appears in any case extremely inefficient, even non-feasible with the performance levels of currently available fuels. However, even only a partial usage of the reaction for braking must be avoided if at all possible for the same reasons.

Another point concerning reaction braking in the region of the atmosphere must additionally be considered—at least for as long as the travel velocity is still of a cosmic magnitude. The exhaust gases, which the vehicle drives ahead of it, would be decelerated more by air drag than the heavier vehicle itself and, therefore, the vehicle would have to travel in the heat of its own gases of combustion.

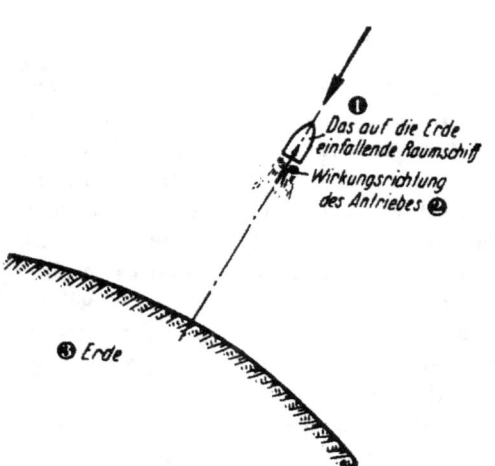

Figure 37. Landing with reaction braking. The descending vehicle is supposed to be "cushioned" by the propulsion system, with the latter functioning opposite to the direction of flight, exactly similar to the ascent "away from the Earth".

Key: 1. The space ship descending to the Earth; 2. Direction of effect of the propulsion system; 3. Earth

The second type of landing, the one using air drag, is brought about by braking the vehicle during its travel through the Earth's atmosphere by means of a parachute or other device (Figure 38). It is critical in this regard that the kinetic energy, which must be removed from the vehicle during this process, is only converted partially into air movement (turbulence) and partially into heat. If now the braking distance is not sufficiently long and consequently the braking period is too short, then the

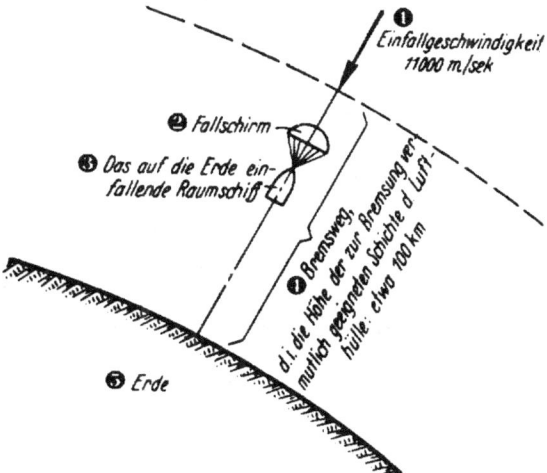

Figure 38. Landing during a vertical descent of the vehicle using air drag braking.

Key: 1. Descent velocity of 11,000 m/sec; 2. Parachute; 3. The space ship descending to Earth; 4. Braking distance, i.e., the altitude of the layers of the atmosphere (approx. 100 km) probably suitable for braking; 5. Earth.

resulting braking heat cannot transition to the environment through conduction and radiation to a sufficient degree, causing the temperature of the braking means (parachute, etc.) to increase continuously.

Now in our case, the vehicle at its entry into the atmosphere has a velocity of around 11,000 meters per second, while that part of the atmosphere having sufficient density for possible braking purposes can hardly be more than 100 km in altitude. According to what was stated earlier, it is fairly clear that an attempt to brake the vehicle by air drag at such high velocities would simply lead to combustion in a relatively very short distance. It would appear, therefore, that the problem of space flight would come to nought if not on the question of the ascent then for sure on the impossibility of a successful return to Earth.

Hohmann's Landing Maneuver

The German engineer Dr. Hohmann deserves the credit for indicating a way out of this dilemma. According to his suggestion, the vehicle will be equipped with wings for landing, similar to an airplane. Furthermore, a tangential (horizontal) velocity component is imparted to the vehicle at the start of the return by means of reaction so that the vehicle does not even impact on the Earth during its descent, but travels around the Earth in such a manner that it approaches within 75 km of the Earth's surface (Figure 39).

This process can be explained in a simple fashion as follows: if a stone is thrown horizontally instead of allowing it to simply drop, then it hits

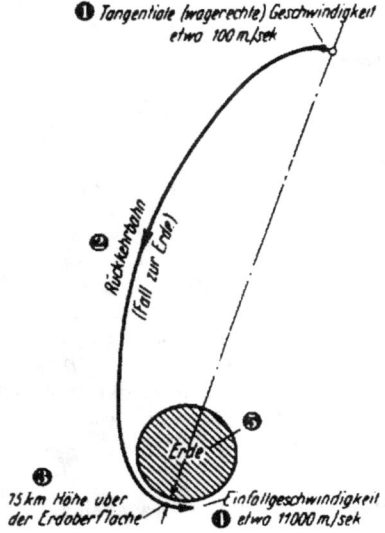

Figure 39. During Hohmann's landing process, the return trajectory is artificially influenced to such an extent that the space ship does not even impact the Earth, but travels around it at an altitude of 75 km.

Key: 1. Tangential (horizontal) velocity of approx. 100 m/sec; 2. Return trajectory (descent to Earth); 3. At an altitude 75 km above the Earth's surface; 4. Descent velocity of approximately 11,000 m/sec; 5. Earth

the ground a certain distance away, and, more specifically, at a greater distance, the greater the horizontal velocity at which it was thrown. If this horizontal velocity could now be arbitrarily increased such that the stone falls not a distance of 10 or 100 meters, not even at distances of 100 or 1,000 km, but only reaches the Earth at a distance of 40,000 km away, then in reality the stone would no longer descend at all because the entire circumference of the Earth measures only 40,000 km. It would then circle the Earth in a free orbit like a tiny moon. However, in order to achieve this from a point on the Earth's surface, the very high horizontal velocity of approximately 8,000 meters per second would have to be imparted to the stone. This velocity, however, becomes that much smaller the further the position from which the object starts is distant from the Earth. At a distance of several hundred thousand km, the velocity is only around 100 meters per second (Figure 39). This can be understood if we visualize that the vehicle gains velocity more and more—solely due to its descent to Earth. According

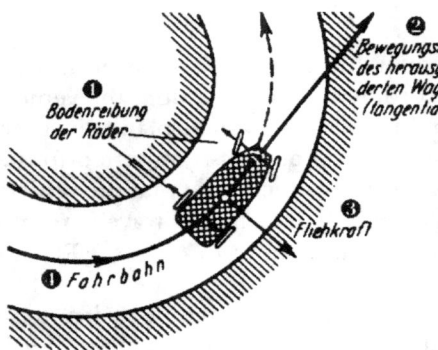

Figure 40. If the centrifugal force becomes extremely large due to excessively rapid travel, it hurls the automobile off the road.

Key: 1. Friction of the wheels on the ground; 2. Direction of motion of the automobile being hurled out (tangential); 3. Centrifugal force; 4. Road.

to what was stated previously, if the descent velocity finally attains the value of 11,000 meters per second, it is then greater by more than 3,000 meters per second than the velocity of exactly 7,850 meters per second that the vehicle would have to have so that it would travel around the Earth (similar to the stone) in a free circular orbit at an altitude of 75 km.

Due to the excessive velocity, the space ship is now more forcefully pushed outward by the centrifugal force than the force of gravity is capable of pulling it inward towards the Earth. This is a process similar, for instance, to that of an automobile driving (too "sharply") through a curve at too high a speed (Figure 40). Exactly as this automobile is hurled outward because the centrifugal force trying to force it off the road is greater than the friction of the wheels trying to keep it on the road, our space ship will—in an analogous way—also strive to exit the free circular orbit in an outward direction and, as a result, to move again away from the Earth (Figure 41).

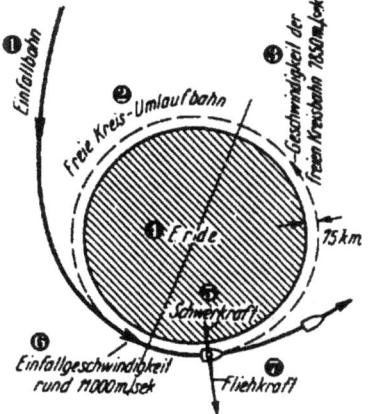

Figure 41. Due to the travel velocity (11,000 instead of 7,850 m/sec!) which is excessive by around 3,000 m/sec, the centrifugal force is greater than the force of gravity, consequently forcing the space ship outward out of the free circular orbit.

Key: 1. Descent trajectory; 2. Free circular orbit; 3. Velocity in the free circular orbit of 7,850 m/sec; 4. Earth; 5. Force of gravity; 6. Descent velocity of around 11,000 m/sec; 7. Centrifugal force.

Landing in a Forced Circular Motion

The situation described above can, however, be prevented through the appropriate use of wings. In the case of a standard airplane, the wings

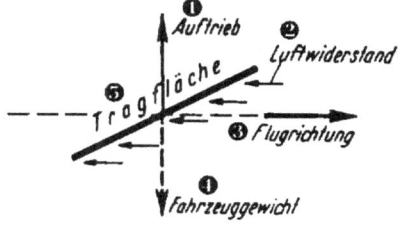

Figure 42. The fundamental operating characteristics of wings during standard heavier-than-air flight: The "lift" caused by air drag is directed upward and, therefore, carries the airplane.

Key: 1. Lift; 2. Air drag; 3. Direction of flight; 4. Weight of the vehicle; 5. Wings; 6. Surface of the Earth.

are pitched upward so that, as a result of the motion of flight, the lift occurs that is supposed to carry the airplane (Figure 42). In our case, the wings are now adjusted in the opposite direction, that is, pitched downward (Figure 43). As a result, a pressure directed downward towards the Earth occurs, exactly offsetting the centrifugal force excess by properly selecting the angle of incidence and in this fashion forcing the vehicle to remain in the circular flight path (Figure 44).

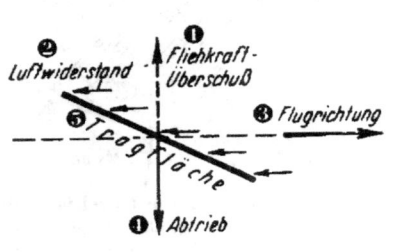

Figure 43. The operating characteristics of wings during the "forced circular motion" of a landing space ship. Here, air drag produces a "negative lift" directed towards the Earth (downward), offsetting the excessive centrifugal force.

Key: 1. Centrifugal force excess; 2. Air drag; 3. Direction of flight; 4. Negative lift; 5. Wings; 6. Surface of the Earth.

For performing this maneuver, the altitude was intentionally selected 75 km above the Earth's surface, because at that altitude the density of air is so thin that the space ship despite its high velocity experiences almost the same air drag as a normal airplane in its customary altitude.

During this "forced circular motion," the travel velocity is continually being decreased due to air drag and, therefore, the centrifugal force excess is being removed more and more. Accordingly, the necessity of assistance from the wings is also lessened until they finally become completely unnecessary as soon as the travel velocity drops to 7,850 meters per second and, therefore, even the centrifugal force excess has ceased to exist. The space ship then circles suspended in a circular orbit around the Earth ("free circular motion," Figure 44).

Since the travel velocity continues to decrease as a result of air drag, the centrifugal force also decreases gradually and accordingly the force of gravity asserts itself more and more. Therefore, the wings must soon become active again and, in particular, acting exactly like the typical airplane (Figure 42): opposing the force of gravity, that is, carrying the weight of the space craft ("gliding flight," Figure 44).

Finally, the centrifugal force for all practical purposes becomes zero with further decreasing velocity and with an increasing approach to the Earth: from now on, the vehicle is only carried by the wings until it finally descends in gliding flight. In this manner, it would be possible to extend the distance through the atmosphere to such an extent that even

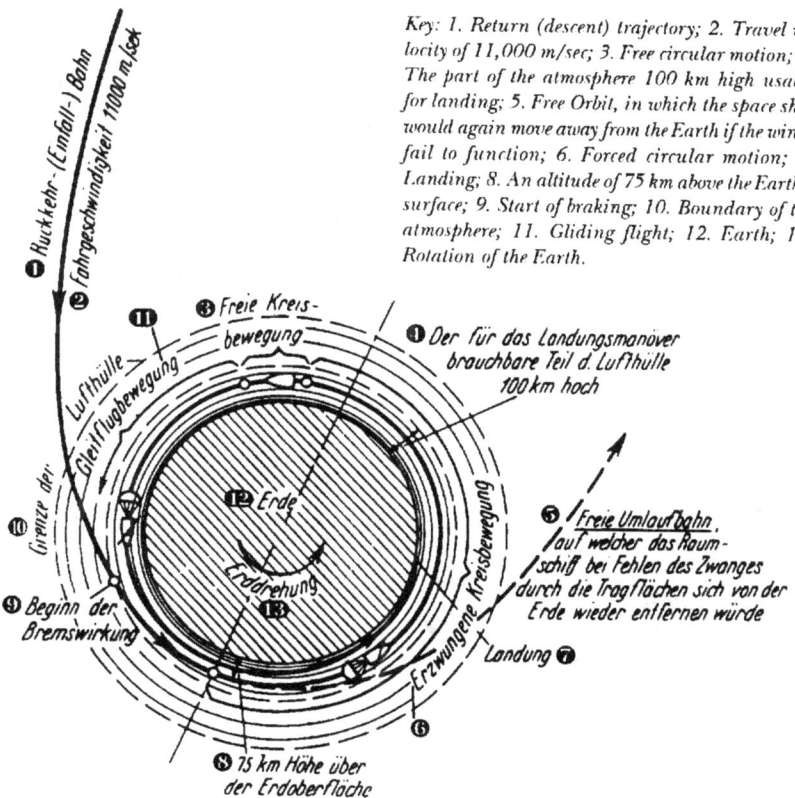

Key: 1. Return (descent) trajectory; 2. Travel velocity of 11,000 m/sec; 3. Free circular motion; 4. The part of the atmosphere 100 km high usable for landing; 5. Free Orbit, in which the space ship would again move away from the Earth if the wings fail to function; 6. Forced circular motion; 7. Landing; 8. An altitude of 75 km above the Earth's surface; 9. Start of braking; 10. Boundary of the atmosphere; 11. Gliding flight; 12. Earth; 13. Rotation of the Earth.

Figure 44. Landing in a "forced circular motion." (The atmosphere and the landing spiral are drawn in the figure—for the purpose of a better overview—higher compared to the Earth than in reality. If it was true to scale, it would have to appear according to the ratios of Figure 8.)

the entire Earth would be orbited several times. During orbiting, however, the velocity of the vehicle could definitely be braked from 11,000 meters per second down to zero partially through the effect of the vehicle's own air drag and its wings and by using trailing parachutes, without having to worry about "overheating." The duration of this landing maneuver would extend over several hours.

Landing in Braking Ellipses

In the method just described, transitioning from the descent orbit into the free circular orbit and the required velocity reduction from 11,000 to 7,850 meters per second occurred during the course of the "forced circular motion." According to another Hohmann recommendation, this can also be achieved by performing so-called "braking ellipses" (Figure 45). In this landing procedure, the wings are not used initially, but braking is performed as vigorously as the previously explained danger of excessive heating will permit by means of a trailing parachute as soon as the vehicle enters into sufficiently dense layers of air.

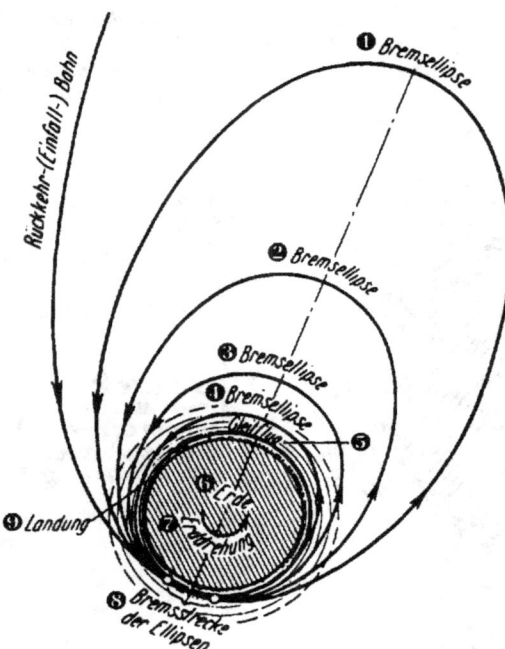

Figure 45. Landing in "braking ellipses." (The atmosphere and landing orbit are drawn here higher than in reality, exactly similar to Figure 44. Reference Figure 8.)

Key: 1. First braking ellipse; 2. Second braking ellipse; 3. Third braking ellipse; 4. Fourth braking ellipse; 5. Glided flight; 6. Earth; 7. Rotation of the Earth; 8. Braking distance of the ellipses; 9. Landing; 10. Return (descent) orbit.

However, the travel velocity, as a result, cannot be decreased to such an extent as would be necessary in order to transition the space ship into free circular motion. An excess of velocity, therefore, still remains and consequently also a centrifugal force that pushes the vehicle outward so that it again exits the atmosphere and moves away from the Earth in a free orbit of an elliptical form (first braking ellipse). The vehicle, however, will not move away to that distance from which it originally started the return flight because its kinetic energy has already decreased during the braking (Figure 45). Due to the effect of gravity, the vehicle will re-return to Earth after some time, again travel through the atmosphere—with a part of its velocity again being absorbed by parachute braking; it will move away from the Earth once again, this time, however, in a smaller elliptical orbit (second braking ellipse), then return again, and so on.

Therefore, narrower and narrower so-called "braking ellipses" will be passed through one after the other corresponding to the progressive velocity decrease, until finally the velocity has dropped to 7,850 meters per second and as a result the free circular motion has been reached. The further course of the landing then occurs with the help of wings in gliding flight, just as in the previously described method. The entire duration of the landing from the initial entry into the atmosphere to the arrival on the Earth's surface is now around 23 hours; it is several times longer than with the method previously described. Therefore, the wings provided anyway for the Hohmann landing will be used to their full extent even at the start and consequently the landing will be performed better in a forced circular motion.

Oberth's Landing Maneuver

The situation is different, however, when wings are not to be used at all, as recommended by Oberth, who also addresses the landing problem in more detail in the second edition of his book. As described above, the first part of the landing is carried out as previously described using braking ellipses (Figure 45), without a need for wings. The subsequent landing process, however, cannot take place in gliding flight because there are no wings. Although the parachute will be inclined with respect to the direction of flight by shortening one side of the shroud, resulting in some lift (an effect similar to that of wings). The use of the propulsion system to a very extensive degree could prove necessary in order to prevent an excessively rapid descent of the vehicle. Therefore, a landing maneuver without the wings could only be achieved at the expense of a fairly significant load of propellants. This assumes that applying reaction braking within the atmosphere would be feasible at all in view of a previously stated danger (a threat due to the vehicle's own gases of combustion). All things considered, the landing according to Hohmann in a "forced circular motion" by means of wings appears, therefore, to represent the most favorable solution.

The Result To Date

We have seen that not only the ascent into outer space but also the assurance of a controlled return to Earth is within the range of technical possibility, so that it does not appear at all justified to dismiss out of hand the problem of space flight as utopia, as people are traditionally inclined to do when they judge superficially. No fundamental obstacles whatsoever exist for space flight, and even those scientific and engineering prerequisites that are available today allow the expectation that this boldest of all human dreams will eventually be fulfilled. Of course, years and decades may pass until this happens, because the technical difficulties yet to be overcome are very significant, and no serious thinking person should fool himself on this point. In many respects, it will probably prove necessary in the practical implementation to alter extensively the recommendations that were proposed to date without a sufficient experimental basis. It will cost money and effort and perhaps even human life. After all, we have experienced all this when conquering the skies! However, as far as technology is concerned, once we had recognized something as correct and possible, then the implementation inevitably followed, even when extensive obstacles had to be overcome—provided, however, that the matter at hand appeared to provide some benefits.

Two Other Important Questions

Therefore, we now want to attempt to show which prospects the result indicated above opens up for the future and to clarify two other existing important questions, because up to this point we have addressed only the technical side of the problem, not its economical and physiological sides. What are the practical and other advantages that we could expect from implementing space travel, and would they be sufficiently meaningful to make all the necessary, and certainly very substantial expenditures appear, in fact, to be beneficial? And, on the other hand, could human life be made possible at all under the completely different physical conditions existing in empty space, and what special precautions would be necessary in this regard?

The answer to these questions will become obvious when we examine in more detail in the following sections the prospective applications of space travel. Usually, one thinks in this context primarily of traveling to distant celestial bodies and walking on them, as has been described in romantic terms by various authors. However, regardless of how attractive this may appear, it would, in any case, only represent the final phase of a successful development of space travel. Initially, however, there would be many applications for space travel that would be easier to implement because they would not require a complete departure from the vicinity of Earth and travel toward alien, unknown worlds.

The Space Rocket in an Inclined Trajectory

For the rocket, the simplest type of a practical application as a means of transportation results when it climbs in an inclined (instead of vertical) direction from the Earth, because it then follows a parabolic trajectory (Figure 46). It is well known that in this case the range is greatest when the ballistic angle (angle of departure)—in our case, the angle of inclination of the direction of ascent—is 45° (Figure 47).

In this type of application, the rocket operates similarly to a projectile, with the following differences, however: a cannon is not necessary to launch it; its weight can be much larger than that of a typical, even very large projectile; the departure acceleration can be selected as small as desired; however, such high departure velocities would be attainable that there would theoretically be no terrestrial limit whatsoever for the ballistic (firing) range of the space rocket.

Therefore, a load could be carried in an extremely short time over very great distances, a fact that could result in the opinion that this method could be used for transporting, for example, urgent freight, perhaps for the post office, telecommunication agency, or similar service organization.

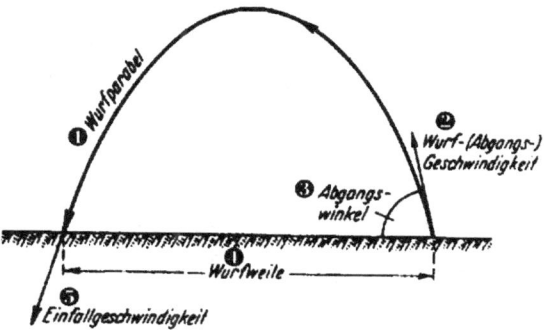

Figure 46. Inclined trajectory.

Key: 1. Parabolic trajectory; 2. Ballistic (departure) velocity; 3. Angle of departure; 4. Range; 5. Impact velocity.

The latter application would, however, only be possible if the descent velocity of the incoming rocket were successfully slowed down to such a degree that the vehicle impacts softly because otherwise it and/or its freight would be destroyed. According to our previous considerations,[16] two braking methods are available in this regard as follows: either by means of reaction or by air drag. Because the former must absolutely be avoided, if at all possible, due to the enormous propellant consumption, only the application of air drag should be considered.

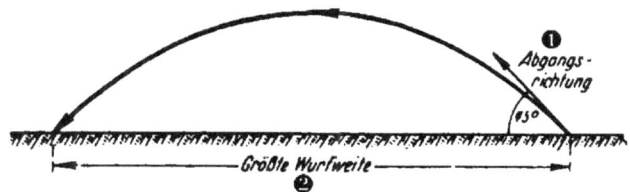

Figure 47. The greatest distance is attained for a given departure velocity when the angle of departure is 45°.

Key: 1. Direction of departure; 2. Greatest distance.

16. See pages 53-61.

Braking could obviously not be achieved with a simple parachute landing, because, considering the magnitudes of possible ranges, the rocket descends to its destination with many times the velocity of a projectile. For this reason, however, the braking distance, which would be available in the atmosphere even in the most favorable case, would be much too short due to the very considerable steepness of the descent. As an additional disadvantage, the main part of the descent velocity would have to be absorbed in the lower, dense layers of air.

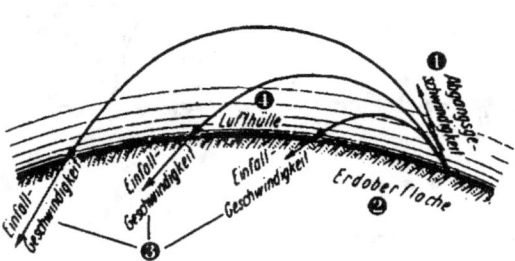

Figure 48. The greater the range, the greater the descent velocity will be (corresponding to the greater departure velocity and altitude necessary for this).

Key: 1. Departure velocity; 2. Earth's surface; 3. Descent velocity; 4. Atmosphere.

This is equally valid even when, as suggested by others, the payload is separated from the rocket before the descent so that it can descend by itself on a parachute, while the empty rocket is abandoned. Neither the magnitude of the descent velocity nor the very dangerous steepness of the descent would be favorably influenced by this procedure.

In order to deliver the freight undamaged to its destination, braking, if it is to be achieved by air drag, could only happen during a sufficiently long, almost horizontal flight in the higher, thin layers of air selected according to the travel velocity—that is, according to Hohmann's landing method (glided landing). Braking would consequently be extended over braking distances not that much shorter than the entire path to be traveled. Therefore, proper ballistic motion would not be realized whatsoever—for the case that braking should occur before the impact—but rather a type of trajectory would result that will be discussed in the next section entitled "The Space Rocket as an Airplane."

With an inclined ballistic trajectory, the rocket could only be used when a "safe landing" is not required, for example, a projectile used in warfare. In the latter case, solid fuels, such as smokeless powder and similar substances, could easily be used for propelling the rockets in the sense of Goddard's suggestion, as has been previously pointed out.[17]

To provide the necessary target accuracy for rocket projectiles of this

---

17. See pages 32-33 and 40.

type is only a question of improving them from a technical standpoint. Moreover, the large targets coming mainly under consideration (such as large enemy cities, industrial areas, etc.) tolerate relatively significant dispersions. If we now consider that when firing rockets in this manner even heavy loads of several tons could safely be carried over vast distances to destinations very far into the enemy's heartland, then we understand what a terrible weapon we would be dealing with. It should also be noted that after all almost no area of the hinterland would be safe from attacks of this nature and there would be no defense against them at all.

Nevertheless, its operational characteristics are probably not as entirely unlimited as might be expected when taking the performance of the rocket propulsion system into consideration, because with a lengthening of the range the velocity also increases at which the accelerated object, in this case the rocket, descends to the target, penetrating the densest layers of air near the Earth's surface (Figure 48). If the range and the related descent velocity are too large, the rocket will be heated due to air drag to such an extent that it is destroyed (melted, detonated) before it reaches the target at all. In a similar way, meteorites falling onto the Earth only rarely reach the ground because they burn up in the atmosphere due to their considerably greater descent velocity, although at a much higher altitudes. In this respect, the Earth's atmosphere would probably provide us at least some partial protection, as it does in several other respects.

No doubt, the simplest application of the rocket just described probably doesn't exactly appear to many as an endorsement for it! Nevertheless, it is the fate of almost all significant accomplishments of technology that they can also be used for destructive purposes. Should, for example, chemistry be viewed as dangerous and its further development as undesirable because it creates the weapons for insidious gas warfare? And the results, which we could expect from a successful development of space rockets, would surpass by far everything that technology was capable of offering to date, as we will recognize in the following discussion.

The Space Rocket as an Airplane

As previously described, Hohmann recommends equipping the space ship with wings for landing. At a certain stage of his landing manoeuver[18], the space ship travels suspended around the Earth in a circular, free orbit ("carried" only by centrifugal force) at an altitude of 75 km and at a corresponding velocity of 7,850 meters per second ("free circular mo-

18. See pages 56-57.

tion," Figure 44). However, because the travel velocity and also the related centrifugal force continually decrease in subsequent orbits, the vehicle becomes heavier and heavier, an effect that the wings must compensate so that the free orbital motion transitions gradually into a glid-

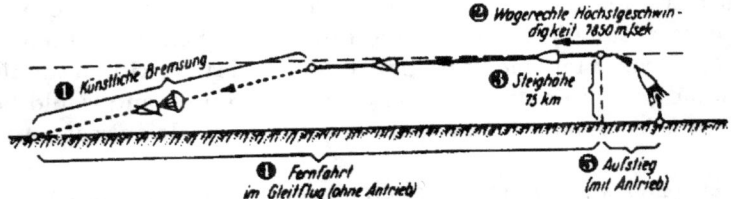

Figure 49. Schematic representation of an "express flight at a cosmic velocity" during which the horizontal velocity is so large (in this case, assumed equal to the velocity of free orbital motion) that the entire long-distance trip can be covered in gliding flight and must still be artificially braked before the landing.

Key: 1. Artificial braking; 2. Highest horizontal velocity of 7,850 m/sec; 3. Altitude of 75 km; 4. Long-distance trip in gliding flight (without power); 5. Ascent (with power).

ing flight. Accordingly, deeper and deeper, denser layers of air will be reached where, in spite of higher drag, the necessary lift at the diminished travel velocity and for the increased weight can be achieved ("gliding motion," Figure 44).

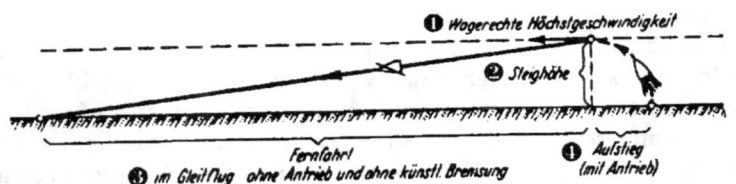

Figure 50. Schematic representation of an "express flight at a cosmic velocity" during which the highest horizontal velocity is just sufficient to be able to cover the entire long-distance trip in gliding flight when any artificial braking is avoided during the flight.

Key: 1. Highest horizontal velocity; 2. Altitude; 3. Long-distance travel in gliding flight without power and without artificial braking; 4. Ascent (with power)

Since even the entire Earth can be orbited in only a few hours in this process, it becomes obvious that in a similar fashion terrestrial express flight transportation can be established at the highest possible, almost cosmic velocities: If an appropriately built space ship equipped with wings climbs only up to an altitude of approximately 75 km and at the same time a horizontal velocity of 7,850 meters per second is imparted to it in the direction of a terrestrial destination (Figure 49), then it could cover the distance to that destination without any further expenditure of energy—in the beginning in an approximately circular free orbit, later more and more in gliding flight and finally just gliding, carried only by atmospheric lift. Some time before the landing, the velocity would finally have to be appropriately decreased through artificial air drag braking, for example, by means of a trailing parachute.

Even though this type of landing may face several difficulties at such high velocities, it could easily be made successful by selecting a smaller horizontal velocity, because less artificial braking would then be necessary. From a certain initial horizontal velocity, even natural braking by the unavoidable air resistance would suffice for this purpose (Figure 50).

In all of these cases, the vehicle requires no power whatsoever during the long-distance trip. If the vehicle is then powered only by a booster rocket—that is, "launched" so to speak by the booster—during the as-

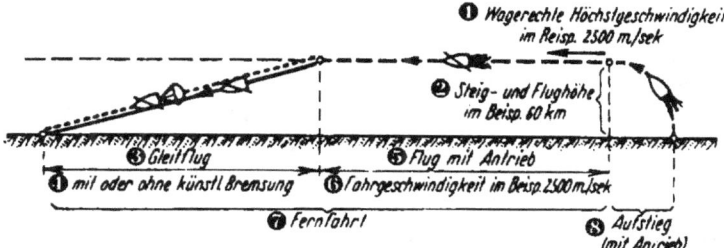

Figure 51. Schematic representation of an "express flight at a cosmic velocity" during which the highest horizontal velocity is not sufficient for covering the entire long-distance trip in gliding flight so that a part of the trip must be traveled under power.

Key: 1. Highest horizontal velocity of 2,500 m/sec in the example; 2. Climb and flight altitude of 60 km in the example; 3. Gliding flight; 4. With or without artificial braking; 5. Power flight; 6. Velocity of travel of 2,500 m/sec in the example; 7. Long-distance trip; 8. Ascent (with power).

cent (until it reaches the required flight altitude and/or the horizontal orbital velocity), then the vehicle could cover the longer path to the destination solely by virtue of its "momentum" (the kinetic energy received) and, therefore, does not need to be equipped with any propulsion equipment whatsoever, possibly with the exception of a small ancillary propulsion system to compensate for possible estimation errors during landing. Of course, instead of a booster rocket, the power could also be supplied in part or entirely by the vehicle itself until the horizontal orbital velocity is attained during the ascent. In the former case, it may be advantageous to let the booster rocket generate mainly the climbing velocity and the vehicle, on the other hand, the horizontal velocity.

In the case of a still smaller horizontal velocity, a certain part of the long-distance trip would also have to be traveled under power (Figure

51). Regardless of how the ascent may take place, it would be necessary in any case that the vehicle also be equipped with a propulsion system and carry as much propellant as is necessary for the duration of the powered flight.

Assuming that benzene and liquid oxygen are used as propellants and thereby an exhaust velocity of 2,500 meters per second is attained, then in accordance with the previously described basic laws of rocket flight technology[19] and for the purpose of attaining maximum efficiency, even the travel velocity (and accordingly the highest horizontal velocity) would have to be just as great during the period when power is being applied, that is, 2,500 meters per second. The optimal flight altitude for this flight would presumably be around 60 km, taking Hohmann's landing procedure into account. At this velocity, especially when the trip occurs opposite to the Earth's rotation (from east to west), the effect of centrifugal force would be so slight that the wings would have to bear almost the entire weight of the vehicle; in that case, the trip would almost be a pure heavier-than-air flight movement rather than celestial motion.

In view of the lack of sufficient technical data, we will at this time refrain from discussing in more detail the design of an aerospace plane powered by reaction (rockets). This will actually be possible—as was indicated previously[20] in connection with the space rocket in general—only when the basic problem of rocket motors is solved in a satisfactory manner for all practical purposes.

On the other hand, the operating characteristics that would have to be used here can already be recognized in substance today. The following supplements the points already discussed about these characteristics: since lifting the vehicle during the ascent to very substantial flight altitudes (35-75 km) would require a not insignificant expenditure of propellants, it appears advisable to avoid intermediate landings in any case. Moreover, this point is reinforced by the fact that breaking up the entire travel distance would make the application of artificial braking necessary to an increasing extent due to the shortening caused as a result of those air distances that can be covered in one flight; these intermediate landings, however, mean a waste of valuable energy, ignoring entirely the losses in time, inconveniences and increasing danger always associated with them. It is inherent in the nature of express flight transportation that it must be demonstrated as being that much more advantageous, the greater (within terrestrial limits, of course) the distances to be covered in one flight, so that these distances will still not be shortened intentionally through intermediate landings.

19. See page 21.
20. See page 51.

Consequently, opening up intermediate filling stations, for example, as has already been recommended for the rocket airplane, among others, by analogy with many projects of transoceanic flight traffic, would be completely counter to the characteristic of the rocket airplane. However, it is surely a false technique to discuss these types of motion by simply taking only as a model the travel technology of our current airplanes, because rocket and propeller vehicles are extremely different in operation, after all.

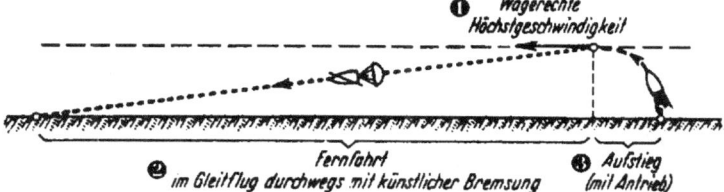

Figure 52. The highest average velocity during the trip is attained when the highest horizontal velocity is selected so large than it can just be slowed if artificial braking is started immediately after attaining that velocity. (In the schematic representations of Figures 49 through 52, the Earth's surface would appear curved in a true representation, exactly as in Figure 53.)

Key: 1. Highest horizontal velocity; 2. Long-distant trip in gliding flight completely with artificial braking; 3. Ascent (with power).

On the other hand, we consider it equally incorrect that rocket airplane travel should proceed not as an actual "flight" at all, but primarily more as a shot (similar to what was discussed in the earlier section), as many authors recommend. Because in this case, a vertical travel velocity component, including the horizontal one, can be slowed down during the descent of the vehicle. Due to the excessively short length of vertical braking distance possible at best in the Earth's atmosphere, this velocity component, however, cannot be nullified by means of air drag, but only through reaction braking. Taking the related large propellant consumption into account, the latter, however, must be avoided if at all possible.

The emergence of a prominent vertical travel velocity component must, for this reason, be inhibited in the first place, and this is accomplished when, as recommended by the author, the trip is covered without exception as a heavier-than-air flight in an approximately horizontal flight path—where possible, chiefly in gliding flight (without power)— that is, proceeding similarly to the last stage of Hohmann's gliding flight landing that, in our case, is started earlier, and in fact, at the highest horizontal velocity.[21]

21. On this point, reference what was described on pages 63-64.

The largest average velocity during the trip, at which a given distance could be traveled in the first place during an express flight of this type, is a function of this distance. The travel velocity is limited by the requirement that braking of the vehicle must still be successful for landing when it is initiated as soon as possible, that is, immediately after attaining the highest horizontal velocity (Figure 52).

The "optimum highest horizontal velocity" for a given distance would be one that just suffices for covering the entire trip in gliding flight to the destination without significant artificial braking (Figures 50 and 53). In the opinion of the author, this represents without a doubt the most advantageous operating characteristics for a rocket airplane. In addition, it is useable for all terrestrial distances, even the farthest, if only the highest horizontal velocity is appropriately selected, primarily since a decreased travel resistance is also achieved at the same time accompanied by an increase of this velocity, because the greater the horizontal velocity becomes, the closer the flight approaches free orbital trajectory around the Earth, and consequently the vehicle loses weight due to a stronger centrifugal force. Also, less lift is necessary by the atmosphere, such that the flight path can now be repositioned to correspondingly higher, thinner layers of air with less drag—also with a lower natural braking effect.

Figure 53. The most advantageous way of implementing an "express flight at a cosmic velocity" is as follows: the highest horizontal velocity is—corresponding to the distance—selected so large ("optimum horizontal velocity") that the entire long-distance trip can be made in gliding flight without power and without artificial braking (see Figure 50 for a diagram).

Key: 1. Gliding flight without power and without artificial braking; 2. Earth's surface; 3. Ascent (with power); 4. "Best case highest horizontal velocity".

The magnitude of the optimum horizontal velocity is solely a function of the length of the distance to be traveled; however, this length can only be specified exactly when the ratios of lift to drag in the higher layers of air are studied at supersonic and cosmic velocities.

However, even smaller highest horizontal velocities, at which a part of the trip would have to be traveled (investigated previously for benzene propulsion), could be considered on occasion. Considerably greater velocities, on the other hand, could hardly be considered because they would make operations very uneconomical due to the necessity of having to destroy artificially, through parachute braking, a significant portion of the energy.

It turns out that these greater velocities are not even necessary! Because when employing the "best case" highest horizontal velocities and even when employing the lower ones, every possible terrestrial distance, even those on the other side of the Earth, could be covered in only a few hours.

In addition to the advantage of a travel velocity of this magnitude, which appears enormous even for today's pampered notions, there is the advantage of the minimal danger with such an express flight, because during the long-distance trip, unanticipated "external dangers" cannot occur at all: that obstacles in the flight path occur is, of course, not possible for all practical purposes, as is the case for every other air vehicle flying at an appropriately high altitude. However, even dangers due to weather, which can occasionally be disastrous for a vehicle of this type, especially during very long-distance trips (e.g., ocean crossings), are completely eliminated during the entire trip for the express airplane, because weather formation is limited only to the lower part of the atmosphere stretching up to about 10 km—the so-called "troposphere." The part of the atmosphere above this altitude—the "stratosphere"—is completely free of weather; express flight transportation would be carried out within this layer. Besides the always constant air streams, there are no longer any atmospheric changes whatsoever in the stratosphere.

Furthermore, if the "optimum velocity" is employed such that neither power nor artificial braking is necessary during the long-distance trip, then the "internal dangers" (ones inherent in the functioning of the vehicle) are reduced to a minimum. Just like external dangers, internal ones can only occur primarily during ascent and landing. As soon as the latter two are mastered at least to that level of safety characteristic for other means of transportation, then express airplanes powered by reaction will not only represent the fastest possible vehicles for our Earth, but also the safest.

Achieving a transportation-engineering success of this magnitude would be something so marvelous that this alone would justify all efforts the implementation of space flight may yet demand. Our notions about terrestrial distances, however, would have to be altered radically if we are to be able to travel, for example, from Berlin to Tokyo or around the entire globe in just under one morning! Only then will we be able to feel like conquerors of our Earth, but at the same time justifiably realizing how small our home planet is in reality, and the longing would increase for those distant worlds familiar to us today only as stars.

The Space Station in Empty Space

Up to this point, we have not even pursued the actual purpose of space ship travel. The goal with this purpose initially in mind would now be as follows: to ascend above the Earth's atmosphere into completely empty space, without having to separate completely from the Earth, however. Solely as a result of this effort, tremendous, entirely new vistas would open up.

Nevertheless, it is not sufficient in this regard to be able only to ascend and to land again. No doubt, it should be possible to perform many scientific observations during the course of the trip, during which the altitude is selected so high that the trip lasts days or weeks. A large-scale use of space flight could not be achieved in this fashion, however, primarily because the necessary equipment for this purpose cannot be hauled aloft in one trip due to its bulk, but only carried one after the other, component-by-component and then assembled at the high altitude.

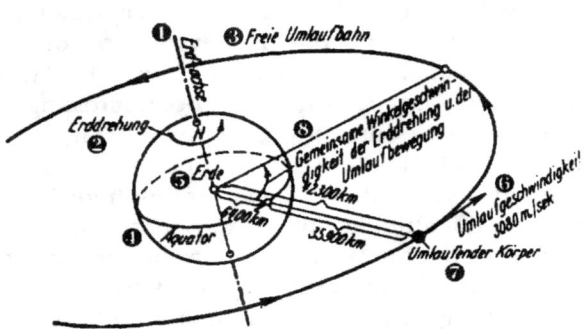

Figure 54. Each object orbiting the Earth in the plane of the equator, 42,300 km from the center of the Earth in a circular orbit, constantly remains freely suspended over the same point on the Earth's surface.

Key: 1. Earth's axis; 2. Earth's rotation; 3. Free orbit; 4. Equator; 5. Earth; 6. Orbital velocity of 3,080 m/sec; 7. Orbiting object; 8. Common angular velocity of the Earth's rotation and of the orbital motion.

The latter, however, assumes the capability of spending time, even arbitrarily long periods, at the attained altitude. This is similar, for instance, to a captive balloon held aloft suspended for long periods without any expenditure of energy, being supported only by the buoyancy of the atmosphere. However, how would this be possible in our case at altitudes extending up into empty space where nothing exists? Even the air for support is missing. And still! Even when no material substance is available, there is nevertheless something available to keep us up there, and in particular something very reliable. It is an entirely natural phenomenon: the frequently discussed centrifugal force.

Introductory paragraphs[22] indicated that humans could escape a heavenly body's gravitational effect not only by reaching the practical limit of gravity, but also by transitioning into a free orbit, because in the latter case the effect of gravity is offset by the emerging forces of inertia (in a circular orbit, solely by the centrifugal force, Figure 5), such that a stable state of suspension exists that would allow us to remain arbitrarily long above the heavenly body in question. Now in the present case, we also would have to make use of this possibility.

Accordingly, it is a matter not only of reaching the desired altitude during the ascent, but also of attaining a given orbital velocity exactly corresponding to the altitude in question (and/or to the distance from the Earth's center). The magnitude of this velocity can be computed exactly from the laws of gravitational motion. Imparting this orbital velocity, which in no case would have to be more than around 8,000 meters per second for the Earth, would present no difficulties, as soon as we have progressed to the point where the completed space vehicle is capable of ascending at that rate.

Among the infinitely large number of possible free orbits around the Earth, the only

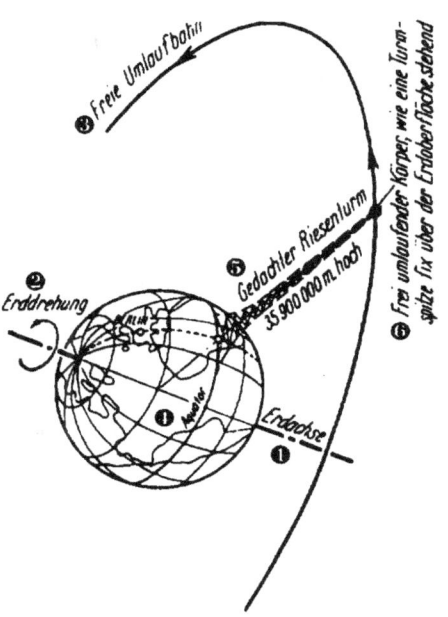

Figure 55. An object orbiting the Earth as in Figure 54 behaves as if it would form the pinnacle of a enormously giant tower (naturally, only imaginary) 35,900,000 meters high.

Key: 1. Earth's axis; 2. Earth's rotation; 3. Free orbit; 4. Equator; 5. Imaginary giant tower 35,900,000 meters high; 6. Freely orbiting object, like a pinnacle of a tower, remaining fixed over the Earth's surface.

ones having significance for our present purpose are approximately circular and of these the only ones of particular interest are those whose radius (distance from the center of the Earth) is 42,300 km (Figure 54). At an assigned orbiting velocity of 3,080 meters per second, this radius corresponds to an orbital angular velocity just as great as the velocity of

22. See pages 6-9.

the Earth's rotation. That simply means that an object circles the Earth just as fast in one of these orbits as the Earth itself rotates: once per day ("stationary orbit").

Furthermore, if we adjust the orbit in such a fashion that it is now exactly in the plane of the equator, then the object would continually remain over one and the same point on the equator, precisely 35,900 km above the Earth's surface, when taking into account the radius of the Earth of around 6,400 km (Figure 54). The object would then so to speak form the pinnacle of a enormously high tower that would not even exist but whose bearing capacity would be replaced by the effect of centrifugal force (Figure 55).

This suspended "pinnacle of the tower" could now be built to any size and equipped appropriately. An edifice of this type would belong firmly to the Earth and even continually remain in a constant position relative to the Earth, and located far above the atmosphere in empty space: a space station at an "altitude of 35,900,000 meters above see level." If this "space station" had been established in the meridian of Berlin, for example, it could continually be seen from Berlin at that position in the sky where the sun is located at noon in the middle of October.

If, instead of over the equator, the space station were to be positioned over another point on the Earth, we could not maintain it in a constant position in relation to the Earth's surface, because it would be necessary in this case to impart to the plane of its orbit an appropriate angle of inclination with respect to the plane of the equator, and, depending on the magnitude of this angle of inclination, this would cause the space station to oscillate more or less deeply during the course of the day from the zenith toward the horizon. This disadvantage could, however, be compensated for in part when not only one but many space stations were built for a given location; with an appropriate selection of the orbital inclination, it would then be possible to ensure that always one of the space stations is located near the zenith of the location in question. Finally, the special case would be possible in which the orbit is adjusted in such a manner that its plane remains either vertical to the plane of the Earth's orbit, as suggested by Oberth, or to that of the equator.

In the same manner, the size (diameter) of the orbit could naturally be selected differently from the present case of a stationary orbit: for example, if the orbit for reasons of energy efficiency is to be established at a greater distance from the Earth (transportation station, see the following) or closer to it, and/or if continually changing the orientation of the space station in relation to the Earth's surface would be especially desired (if necessary, for a space mirror, mapping, etc, see the following).

What would life be like in a space station, what objectives could the station serve and consequently how would it have to be furnished and equipped? The special physical conditions existing in outer space, weightlessness and vacuum, are critical for these questions.

The Nature of Gravity and How it can be Influenced

At the beginning of this book we discussed[23] the so-called inertial forces and we distinguished several types of these forces: gravity, inertia and, as a special case of the latter, centrifugal force. At this point, we must concern ourselves in somewhat more detail with their nature.

It is the nature of these forces that they do not act only upon individual points of the surface of the object like other mechanical forces, but that they act simultaneously on all points even its internal ones. Since this special characteristic feature is common to all inertial forces, it is, therefore, entirely immaterial as far as a practical effect is concerned what type of inertial force is involved. It will always affect an object in the same fashion, as the force of gravity, and we will likewise feel it in every case as the well-known "weighty feeling," regardless of whether the force is gravity, inertia, centrifugal force or even the result of several of these forces. As a result of this complete uniformity of effect, it is possible that different types of inertial

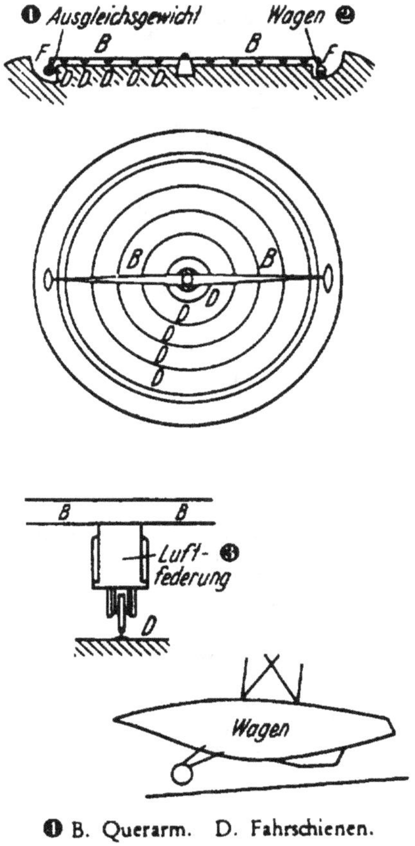

23. See pages 3-5.

❶ B. Querarm. D. Fahrschienen.

Figure 56. Carousel, accordance to Oberth. This equipment and that shown in Figure 57 are both designed to produce artificially the condition of an increased force of gravity for the purpose of carrying-out physiological experiments.

Key: 1. Counterbalancing weight; 2. Vehicle; 3. Pneumatic cushioning; 4B. Lateral arm; 4D. Tracks.

forces can mutually strengthen or weaken or also completely cancel each other.

We are already familiar with an example of the occurrence of a mutual strengthening of inertial forces when studying the ascent of space rockets.[24] In this case, the force of gravity is increased due to the resulting inertia as long as there is thrust, something that makes itself felt for all practical purposes like a temporary increase of the force of gravity (Figure 22).

However, even under normal terrestrial conditions, the state of an increased force of gravity—and even for any desired duration—can be produced, when the centrifugal force is used for this purpose. Technical applications include, for example, different types of centrifuges. Their principle could be applied even on a large scale using a carrousel built especially for this purpose (Figure 56) or, better yet, in specially-built giant centrifuges (Figures 57 and 58). At an appropriately high rate of rotation, a very significant multiplication of the gravitational effect would be achievable in this fashion.

On the other hand, a longer lasting decrease or cancellation of gravity (that is, generating a continuous weightless state) is not possible under terrestrial conditions, because—to emphasize this once again—the force of gravity cannot be eliminated in any other way whatsoever than through the

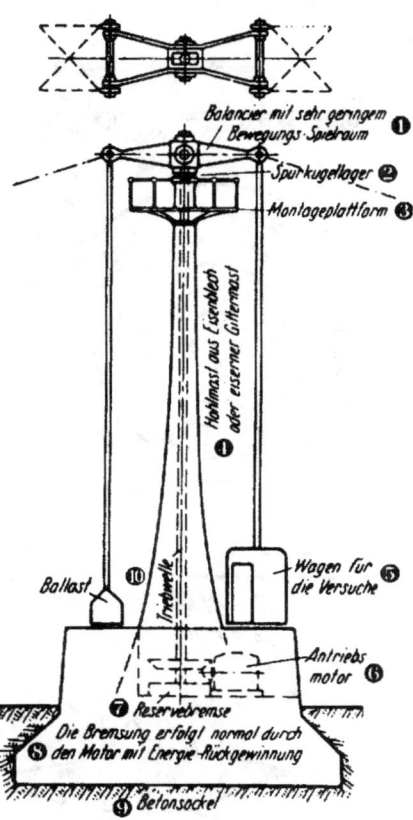

Figure 57. Giant centrifuge according to the author's recommendation. This equipment and that shown in Figure 56 are both designed to produce artificially the condition of an increased force of gravity for the purpose of carrying out physiological experiments.

Key: 1. Beam with a slight clearance of motion; 2. Ball bearing; 3. Maintenance platform; 4. Tubular pole made from sheet iron or iron lattice tower; 5. Gondola for the experiments; 6. Drive motor; 7. Back-up brake; 8. Braking occurs normally by the motor using energy recovery; 9. Concrete base; 10. Drive shaft.

24. See pages 29-31.

opposition of another inertial force of the same magnitude. Therefore, an object can be prevented by supports from falling (i.e., responding to the force of gravity). Its weight, however, cannot be cancelled, a point proven by the continual presence of its pressure on the support. Any experiment to remove the influence of the force of gravity from an object, for instance, by some change of its material structure, would, no doubt, be condemned to failure for all times.

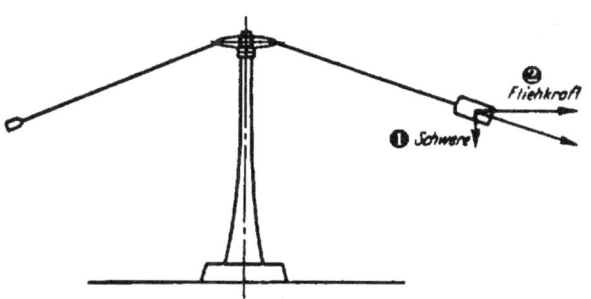

Figure 58. The giant centrifuge in operation.

Key: 1. Gravity; 2. Centrifugal force.

On the Earth's surface neither a correspondingly strong different force of gravity is available nor can centrifugal forces be generated in an object in such a way that it is transposed into an observable weightless state as a result of their effect.

It is, however, possible on the Earth—if only for a short duration—to offset the force of gravity through the third inertial force, the force of inertia. Every day, we can experience this type of occurrence of weightlessness on ourselves or observe it on other objects, namely in the free fall state. That an object falls means nothing more than that it is moved towards the center of the Earth by its weight, and, more specifically, at an acceleration (of 9.81 m/sec^2, a value familiar to us) that is exactly so large that the force of inertia activated in the object as a result exactly cancels the object's weight (Figure 59), because if a part of this weight still remained, then it would result in a corresponding increase of the acceleration and consequently of the inertia (opposing gravity in this case).

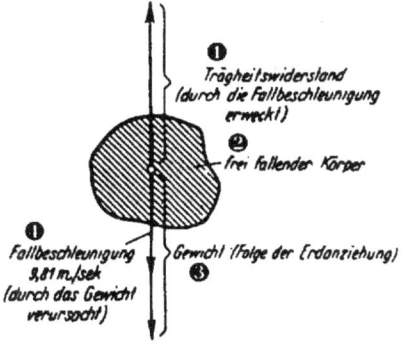

Figure 59. The interplay of forces on a free falling object.

Key: 1. Inertia (activated by the acceleration due to gravity); 2. Free falling object; 3. Weight (as a result of the Earth's attraction); 4. Acceleration due to gravity of 9.81 m/sec^2 (caused by the weight).

In the free fall or during a jump, we are weightless according to this reasoning. The sensation that we experience during the fall or jump is that of weightlessness; the behavior we observe in an object during free fall would be the same in a weightless state generated in another way. Since, however, falling can only last moments if it is not supposed to lead to destruction (the longest times are experienced during parachute jumping, ski jumping, etc.), the occurrence of the weightless state on Earth is possible for only a very short time. Nevertheless, Oberth was successful in conducting very interesting experiments in this manner, from which conclusions can be made about the behavior of various objects and about the course of natural phenomena in the weight-free state.

Completely different, however, are the conditions during space travel. Not only can free fall last for days and weeks during space travel. It would also be possible to remove permanently the effect of gravity from an object: more specifically and as already stated in the introductory chapter[25] by using the action of inertial forces produced by free orbital motion, in particular, of the centrifugal force. As has been previously stated, the space station makes use of this. An orbiting station is in the state of complete freedom from gravity lasting indefinitely ("a stable state of suspension").

The Effect of Weightlessness on the Human Organism

How does the absence of gravity affect the human organism? The experience during free fall shows that a state of weightlessness lasting only a short time is not dangerous to one's health. Whether this would be true in the case of long-lasting weightlessness, however, cannot be predicted with certainty because this condition has not been experienced by anyone. Nevertheless, it may be assumed with a high probability, at least in a physiological sense, because all bodily functions occur through muscular or osmotic forces not requiring the help of gravity. Actually, all vital processes of the body have been shown to be completely independent of the orientation of the body and function just as well in a standing, a prone, or any other position of the body.

Only during very long periods in a weightless state could some injury be experienced, perhaps by the fact that important muscle groups would atrophy due to continual lack of use and, therefore, fail in their function when life is again operating under normal gravitational conditions (e.g, following the return to Earth). However, it is probable that these effects could be counteracted successfully by systematic muscular exercises; be-

---

25. See pages 1, 3.

sides, it might be possible to make allowance for these conditions by means of appropriate technical precautions, as we will see later.

Apparently, the only organ affected by the absence of gravity is the organ of equilibrium in the inner ear. However, it will no longer be required in the same sense as usual, because the concept of equilibrium after all ceases to exist in the weightless state. In every position of the body, we have then the same feeling: "up" and "down" lose their usual meaning (related to the environment); floor, ceiling and walls of a room are no longer different from one another.

However, in the beginning at least, the impression of this entirely unusual condition may cause a strongly negative psychological effect. Added to this is the effect that is directly exercised on the nervous system by the weightless state. The most important sensations related to this effect are as follows: the previously discussed effect on the organ of equilibrium, cessation of the perception of a supporting pressure against the body, and certain changes in the feelings in the muscles and joints.

However, this complex of feelings is known to us so far only from the free fall state because, as already discussed, we can experience freedom from gravity under terrestrial conditions only during falling; involuntarily, we will, therefore, feel anxiety related to the falling, as well as other psychological states aroused by this unusual situation during a cessation of the feeling of gravity, when the lack of gravity is not even caused by falling, but in another way (such as, in the space station by the effect of centrifugal force).

In any event, it can be expected based on previous experiences (pilots, ski jumpers, etc.) that it will be possible through adaptation to be able easily to tolerate the weightless state even in a psychological sense. Adapting occurs that much sooner, the more one is familiar with the fact that "weightless" and "falling" need not be related to one another. It can even be assumed that anxiety is altogether absent during a gradual release from the feeling of gravity.

Oberth has addressed all of these issues in depth. By evaluating his results, they can be summarized as follows: while weightlessness could certainly be tolerated over a long time, although perhaps not indefinitely, without significant harm in a physical sense, this cannot be stated with certainty in a psychological sense, but can be assumed as probable none the less. The course of the psychological impressions apparently would more or less be the following: in the beginning—at least during a rapid, abrupt occurrence of the absence of gravity—anxiety; the brain and senses are functioning extremely intensively, all thoughts are strongly factual and are quickly comprehended with a penetrating logic; time appears to move more slowly; and a unique insensitivity to pains and feelings of displeasure appear. Later, these phenomena subside, and only a certain feeling of elevated vitality and physical fitness remain, perhaps similar to that experienced after taking a stimulant; until finally after a longer period of adaptation, the psychological state possibly becomes entirely normal.

The Physical Behavior of Objects when Gravity is Missing

In order to be able to form a concept of the general physical conditions existing in a weightless state, the following must be noted: the force

Figure 60. A room of the space station in which a weightless state exists and which is furnished accordingly: the walls are completely cushioned and equipped with straps. No loose object is present.

K—Lockable small chests for holding tools and similar items.
L —Openings for admitting light (reference Page 96).
O—Openings for ventilation (reference Page 97).
z —Movement of people by pulling.
a —Movement of people by pushing off.

Key: 1. Direction of motion.

of the Earth's gravity pulling all masses down to the ground and thus ordering them according to a certain regularity is no longer active. Accordingly almost following only the laws of inertia (inertial moment), bodies are moving continually in a straight line in their momentary direction of motion as long as no resistance impedes them, and they react solely to the forces (molecular, electrical, magnetic, mass-attracting and others) acting among and inside themselves. These unusual conditions must, however, lead to the result that all bodies show a completely altered behavior and that, in accordance with this behavior, our unique actions and inactions will develop in a manner entirely different from previous ones.

Figure 61. Writing in the weightless state: for this purpose, we have to be secured to the tabletop, for example, by means of leather straps (G) in order to remain at the table at all (without having to hold on). A man floats in from the next room through the (in this case, round) door opening, bringing something with him.

Therefore, human movement can now no longer occur by "walking." The legs have lost their usual function. In the absence of the pressure of weight, friction is missing under the soles; the latter stick, therefore, considerably less to the ground than even to the smoothest patch of ice. To move, we must either pull ourselves along an area with our hands (Figure 60, z), for which purpose the walls of the space station would have to be furnished with appropriate handles (for instance, straps similar to those of street cars) (Figures 60 and 61), or push ourselves off in the direction of the destination and float towards it (Figure 60, a).

It will probably be difficult for the novice to maintain an appropriate control over his bodily forces. This, however, is necessary: since he impacts the opposite wall of the room with the full force of pushing off, too much zeal in this case can lead very easily to painful bumps. For this reason, the walls and in particular all corners and edges would have to be very well cushioned in all rooms used by human beings (Figure 60).

Pushing off can also be life threatening, more specifically, when it occurs not in an enclosed room but in the open; e.g., during a stay (in the space suit, see the following) outside of the space station, because if we neglected to take appropriate precautionary measures in this case and missed our destination while pushing off, then we would continually float further without end into the deadly vacuum of outer space. The no less terrible possibility of "floating off into space" now threatens as a counterpart to the terrestrial danger of "falling into the depths." The saying "man overboard" is also valid when gravity is missing, however in another sense.

Since bodies are now no longer pressed down upon their support by their weight, it, of course, has no purpose that an item is "hung up" or "laid down" at any place, unless it would stick to its support or would be held down by magnetic or other forces. An object can now only be stored by attaching it somewhere, or better yet locking it up. Therefore, the rooms of the space station would have to be furnished with reliably lockable small chests conveniently placed on the walls (Figures 60 and 61, K).

Clothes racks, shelves and similar items, even tables, as far as they are meant to hold objects, have become useless pieces of furniture. Even chairs, benches and beds can no longer satisfy their function; humans will have to be tied to them in order not to float away from them into any corner of the room during the smallest movement. Without gravity, there is neither a "standing" nor a "sitting" or "lying." In order to work, it is, therefore, necessary to be secured to the location of the activity: for example, to the table when writing or drawing (Figure 61). To sleep, we do not have to lie down first, however; we can take a rest in any bodily position or at any location in the room.

However, despite this irregularity in the physical behavior of freely moving objects caused by the absence of gravity, the manner is actually not completely arbitrary as to how these objects now come to rest. The general law of mass attraction is valid even for the space station itself and causes all masses to be attracted toward the common center of mass; however—due to the relative insignificance of the entire mass—they are attracted at such an extremely slight acceleration that traveling only one meter takes hours. However, non-secured objects will finally impact one of the walls of the room either as a result of this or of their other random movement, and either immediately remain on this wall or, if their velocity was sufficiently large, bounce back again and again among the walls of the room depending on the degree of elasticity, floating back and forth until their energy of movement is gradually expended and they also come to rest on one of the walls. Therefore, all objects freely suspended within the space station will land on the walls over time; more specifically, they will approach as close as possible to the common center of mass of the structure.

This phenomenon can extend over hours, sometimes over many days, and even a weak air draft would suffice to interfere with it and/or to tear objects away from the wall, where they are already at rest but only adhering very weakly, and to mix them all up. Consequently, there is, practically-speaking, no regularity to the type of motion of weightless masses.

The latter is especially unpleasant when objects are in one room in significant numbers. If these objects are dust particles, they can be collected and removed in a relatively easy manner by filtering the air with vacuum cleaners or similar devices. However, if they are somewhat bigger as, for example, through the careless emptying of a sack of apples into a room, then the only alternative would be trapping them by means of nets. All objects must be kept in a safe place, because the ordering power of gravity now no longer exists: matter is "unleashed."

Also, clothing materials are on strike, because they no longer "fall," even if they were made of a heavy weave. Therefore, coats, skirts, aprons and similar articles of clothing are useless. During body movements, they would lay totally irregularly in all possible directions.

The behavior of liquids is especially unique in a weightless state. As is well known, they try under normal conditions to attain the lowest possible positions, consequently obeying gravity by always clinging completely to the respective supports (to the container, to the ground, etc.). If gravity is missing, however, the individual particles of mass can obey their molecular forces unimpeded and arrange themselves according to their characteristics.

In the weightless state therefore, liquids take on an independent shape, more specifically, the simplest geometric shape of an object: that of a ball. A prerequisite for this is, however, that they are subjected to only their forces of cohesion; that is, they are not touching any object they can "moisten." It now becomes understandable why water forms drops when falling. In this state, water is weightless, according to what has been previously stated; it takes on the shape of a ball that is distorted to the form of a drop by the resistance of air.

However, if the liquid is touching an object by moistening it, then overwhelmingly strong forces of cohesion and adhesion appear. The liquid will then strive to obey these forces, spreading out as much as possible over the surface of the object and coating it with a more or less thick layer. Accordingly for example, water in only a partially filled bottle will not occupy the bottom of the bottle, but, leaving the center empty, attempts to spread out over all the walls of the container (Figure 62). On the other hand, mercury, which is not a moistening liquid, coalesces to a ball and adheres to one wall of the container, remaining suspended in the bottle (Figure 63).

In both instances, the position of the body is completely immaterial. Therefore, the bottle cannot be emptied by simply tilting it, as is usually the case. To achieve this effect, the bottle must either be pulled back rapidly (accelerated backwards, Figure 64) or pushed forward in the direction of the outlet and/or then suddenly halted in an existing forward motion (slowing it down in a forward movement, also as in Figure 64), or finally swung around in a circle (Figure 65).

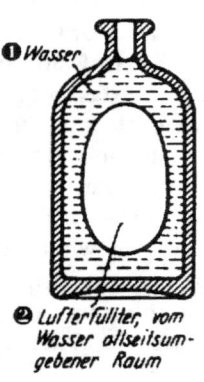

Figure 62. Dispersion of water in only a partially filled bottle in the absence of gravity.

Key: 1. Water; 2. An air-filled space surrounded on all sides by water.

Figure 63. Behavior of mercury in a bottle in the absence of gravity.

Key: 1. Ball of mercury.

The liquid will then escape out of the bottle as a result of its power of inertia (manifested in the last case as centrifugal force), while taking in air at the same time (like gurgling when emptying the bottle in the usual fashion). A prerequisite for this, however, is that the neck of the bottle is sufficiently wide and/or the motion is performed with sufficient force that this entry of air can actually take place against the simultaneous outward flow of water.

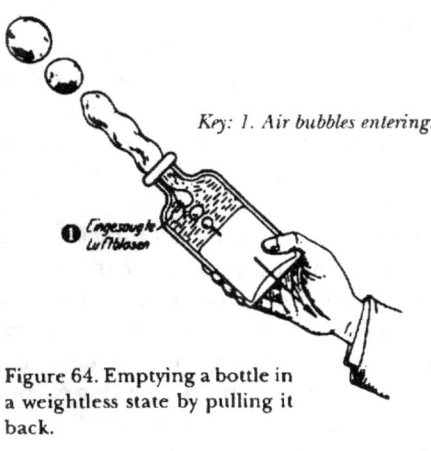

Key: 1. Air bubbles entering.

Figure 64. Emptying a bottle in a weightless state by pulling it back.

[It is interesting to note that strictly speaking the described method of emptying a bottle in the absence of gravity by pulling it back or halting it proceeds in reality as if the water is poured out by turning the bottle upside down in the presence of gravity. Of course, these are completely analogous to physical phenomena (on Earth), if the motion of pulling back and/or halting is performed exactly at the acceleration of gravity (9.81 m/sec^2 for us), because as is known in accordance with the general theory of relativity, a system engaged in accelerated or decelerated motion is completely analogous to a gravitational field of the same acceleration. In the case of the described method of emptying, it can be stated that the forces of inertial mass that are activated by pulling back or stopping of the system operate in place of the missing gravity, including the bottle and its contents.]*

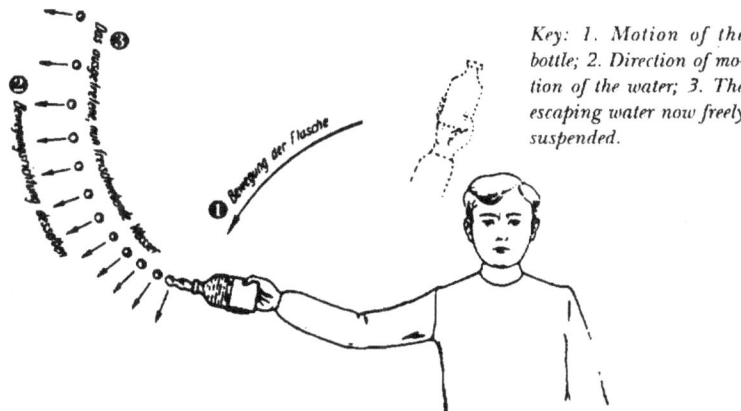

Figure 65. In the absence of gravity, swinging a bottle of water in a circle in order to empty it. (In reality, the escaping liquid will probably not be dispersed in such a regular fashion as the discharge curve indicates.)

Key: 1. Motion of the bottle; 2. Direction of motion of the water; 3. The escaping water now freely suspended.

After escaping from the bottle, the liquid coalesces into one or more balls and will continue floating in the room and may appear similar to soap bubbles moving through the air. Finally, every floating liquid ball of this type must then impact on one of the

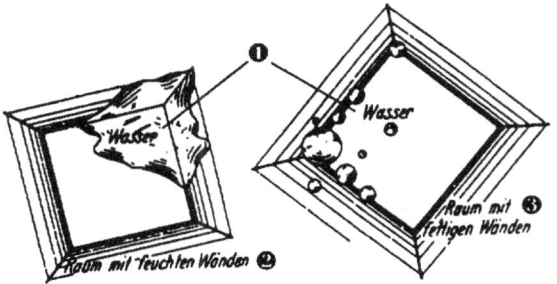

Figure 66. In the absence of gravity, escaping water would spread out over the walls in a room whose walls are easily moistened (e.g., they are somewhat damp; diagram on the left); in a room whose wall are not easily moistened (e.g., one coated with oil), the water coalesces into balls and adheres to the walls (diagram on the right). See end of 1st paragraph, p.86.

* Author's square brackets.

Key: 1. Water; 2. Room with damp walls; 3. Room with walls coated with oil.

walls of the room. If it can moisten one of those walls, then it will try to spread out over them (left portion of Figure 66).

Otherwise as a result of the push, the liquid will scatter into numerous smaller balls, somewhat similar to an impacting drop of mercury. These balls float away along the walls or perhaps occasionally freely through the room, partially coalescing again or scattering once again until their kinetic energy has finally been expended and the entire amount of liquid comes to rest, coalesced into one or more balls adhering to the walls (right portion of figure 66). (In this regard, compare the previous statements about the phenomena in a bottle, Figures 62 and 63.)

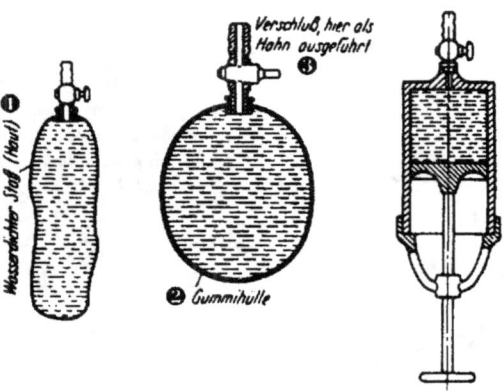

Figure 67. In the absence of gravity, the otherwise usual liquid containers are replaced by sealable flexible tubes (left diagram), rubber balloons (center diagram) or syringe-type containers (right diagram).

Key: 1. Waterproof material (skin); 2. Rubber container; 3. Stopper functioning as a spigot here.

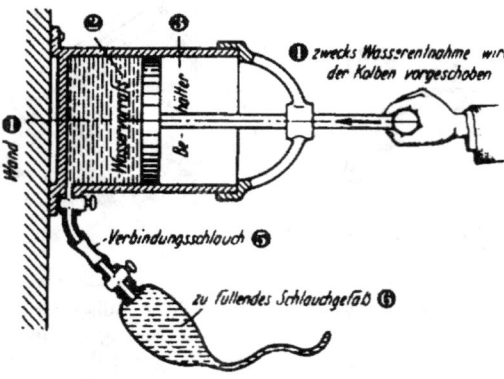

Figure 68. Filling a water vessel in the weightless state.

Key: 1. Wall; 2. Water supply; 3. Container; 4. The plunger is pushed forward for the purpose of removing water; 5. Connecting tube; 6. Tubular container being filled.

Given this unusual behavior of the liquid, none of the typical containers, such as bottles, drinking glasses, cooking pots, jugs, sinks, etc., could be used. It would hardly be possible to fill them. However, even if, by way of example, a bath could be prepared, we would not be able to take it because in the shortest time and to our disappointment, the water would have spread out of the bathtub over the walls of the room or adhered to them as balls.

For storing liquids, only sealable flexible tubes, rubber balloons or containers with plunger-like, adjustable bottoms, similar to syringes, would be suitable (Figure 67), because only items of this nature can be filled (Figure 68) as well as easily emptied. Containers with plunger-like, adjustable bottoms function by pressing together the sides or by advancing the plunger to force out the contents (Figure 69). In the case of elastic balloons, which are filled by expanding them, their tension alone suffices to cause the liquid to flow out when the spigot is opened (Figure 70). These types of pressure-activated containers (fitted with an appropriate mouth piece) would now have to be used for drinking in place of the otherwise typical, but now unusable drinking vessels.

Similarly, the various eating utensils, such as dishes, bowls, spoons, etc., can no longer be used. If we made a careless move, we would have to float through the room chasing after their perhaps savory contents. Eating would, therefore, be possible only in two principal ways: either by eating the food in a solid form, such as bread, or drinking it in a liquid or mushy state using the pressure activated containers described above. The cook would have to deliver the food prepared in this manner.

In his important activity, the cook would be faced with particularly significant problems,

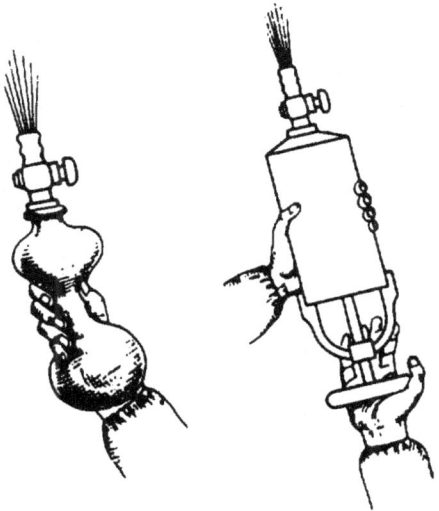

Figure 69. In the absence of gravity, emptying a liquid container can be accomplished in an expedient manner only by pushing out (pressing out) the contents.

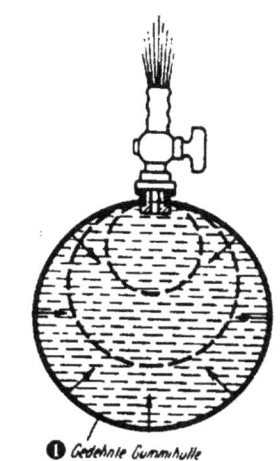

Figure 70. In the case of elastic rubber balloons filled under pressure, the contents flow out of their own accord when the spigot is opened.

Key: 1. Expanded rubber container.

but they can also be overcome. The cook could use, for example, sealable electrical cooking appliances, constantly rotating when in use, so that (instead of the now missing gravity) the generated centrifugal force presses the contents against the walls of the container; there would also be other possibilities. In any case, cooking would not be easy, but certainly possible, as would eating and drinking. Washing and bathing as we know them would have to be completely dropped, however! Cleaning up could only be accomplished by rubbing with damp towels, sponges or the like lathered according to need, accepting whatever success this method would achieve.

The more in depth we consider the situation, the more we must recognize that in reality it would in no way be an entirely unblemished pleasure to be able to float like angels, freed from all bothersome weight; not even if this state of weightlessness were perceived as pleasant. Because, gravity not only holds us in her grip; it also forces all other objects to the ground and inhibits them from moving chaotically, without regularity, freely left to chance. It is perhaps the most important force imposing order upon our existence. Where gravity is absent, everything is in the truest sense "standing on its head," having lost its foothold.

Without Air

Human life can exist only in the presence of appropriately composed gaseous air: on the one hand, because life is a combustion phenomenon and, therefore, requires for its maintenance a permanent supply of oxygen, which the human organism, however, can only obtain from gaseous air by breathing; and, on the other hand, because the body must always be surrounded by a certain pressure, without which its water content would vaporize and the vessels would burst. It is necessary to provide a man-made supply of air if our terrestrial life is to be maintained in empty space.

To accomplish this, people in empty space must always be completely surrounded by absolutely airtight enclosures, because only within such capsules can the air be artificially maintained at the appropriate pressure and in the correct composition by automatic equipment.

Actually, we are only concerned with larger enclosed spaces extending from the size of a closet up to the size of an entire building, because only the latter would be possible for a longer stay. The walls of these structures would have to be built in accordance with the fundamentals of steam boiler construction because they have to withstand an internal air pressure (relative to empty space) of 1 atmosphere; they should not only have an appropriate strength but also curved surfaces if at all possible, because flat ones require special braces or supports in view of the overpressure. The nitrogen necessary for the air, and especially the oxygen, would always have to be maintained in sufficient supply in the liquid state in their own tanks through continual resupply from Earth.

However, in order to exist also outside of enclosed capsules of this type in empty space, airtight suits would have to be used, whose interior is also supplied automatically with air by attached devices. Such suits would be quite similar to the familiar underwater diving suits. We will call them "space suits." The subject of space suits will be addressed in more detail later.

It can be seen that we are dealing here with problems similar to those of remaining under water, that is, with submarine technology and diving practices. On the basis of the extensive experiences already gathered there on the question of supplying air artificially, it can be stated that this problem, without question, is entirely solvable also for a stay in empty space.

Perpetual Silence Prevails in Empty Space

Air not only has direct value for life. Indirectly, it also has an important significance because to a far-reaching extent it influences natural phenomena that are extremely important for the functional activities of life: heat, light and sound.

Sound is a vibrational process of air and can, therefore, never exist in the absence of the latter. For this reason, a perpetual silence exists in empty space. The heaviest cannon could not be heard when fired, not even in its immediate vicinity. Normal voice communication would be impossible. Of course, this does not apply for the enclosed, pressurized rooms, within which the same atmospheric conditions will be maintained artificially as on the Earth's surface; it is true, however, outside of these rooms (in the space suit). There, voice communication would only be possible via telephones.

Sunshine During Nighttime Darkness

Even the lighting conditions are considerably altered in space. As is generally known, the concept of day is associated with the notion of a blue sky or sunlit clouds and scattering of light in all directions, without direct sunlight being necessary. All of these phenomena are, however, due only to the presence of the Earth's atmosphere, because in it a part of the incident radiation of the sun is refracted, reflected and scattered in all directions many times; one of the results of this process is the impression of a blue color in the sky. The atmosphere produces a widespread and pleasant, gradual transition between the harshness of sunlight and darkness.

This is all impossible in empty space because air is absent there. As a result, even the concept of day is no longer valid, strictly speaking. Without letup, the sky appears as the darkest black, from which the infinite number of stars shine with extreme brightness and with a constant untwinkling light, and from which the sun radiates, overwhelming everything with an unimaginably blinding force.

And yet as soon as we turn our gaze from it, we have the impression of night, even though our back is being flooded by sunlight because, while the side of the object (e.g., an umbrella) turned towards the sun is brightly illuminated by its rays, nighttime darkness exists on the back side. Not really complete darkness! After all, the stars shine from all sides and even the Earth or Moon, as a result of their reflectivity, light up the side of the object in the sun's shadow. But even in this case, we observe only the harshest, brightest light, never a mild, diffuse light.

Unlimited Visibility

In one regard however, the absence of air also has advantages for lighting conditions in empty space. After all, it is generally known what great effect the property of air exerts on visibility (e.g., in the mountains, at sea, etc.), because even on clear days, a portion of the light is always lost in the air, or rather through small dust and mist particles constantly suspended in it.

The latter effect is, however, very disadvantageous for all types of long range observations, especially those of astronomy. For this reason, observatories are built if at all possible at high altitudes on mountains because there the air is relatively the clearest. However, there are limits. Furthermore, the flickering of fixed stars, likewise a phenomenon caused only by the presence of air, cannot be avoided even at these high locations. Neither is it possible to eliminate the scattered light (the blue of the sky), which is very bothersome for astronomical observations during the day and is caused also by the atmosphere, thus making it very difficult to investigate those heavenly bodies that cannot be seen during complete darkness, such as Mercury, Venus, and, not least of all, the sun itself.

All of these adverse conditions are eliminated in the empty space of the universe; here, nothing weakens the luminosity of the stars; the fixed stars no longer flicker; and the blue of the sky no longer interferes with the observations. At any time, the same favorable, almost unlimited possibilities exist, because telescopes of any arbitrary size, even very large ones, could be used since optical obstructions no longer exist.

Without Heat

Especially significant is the effect the absence of air exerts on the thermal conditions of outer space. Because as we know today heat is nothing more than a given state of motion of the smallest material particles of which the materials of objects are composed, its occurrence is always associated with the supposition that materials exist in the first place. Where these materials are missing, heat cannot, therefore, exist: empty space is "heatless" for all practical purposes. Whether this is completely correct from a theoretical standpoint depends on the actual validity of the view expressed by some experts that outer space is filled with a real material, distributed very finely, however. If a total material emptiness exists, then the concept of temperature loses its meaning completely.

This view does not contradict the fact that outer space is permeated to a very high degree by the sun's thermal rays and those of the other fixed stars, because the thermal rays themselves are not equivalent with heat! They are nothing more than electromagnetic ether waves of the same type as, for example, light or radio waves; however, they have a special property in that they can generate, as soon as they impact some material, the molecular movement that we call heat. But this can only happen when the waves are absorbed (destroyed) by the affected materials during the impacting, because only in this case is their energy transmitted to the object and converted into the object's heat.

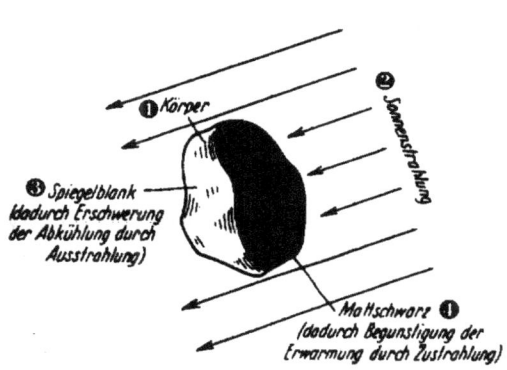

Figure 71. Heating of an object in empty space by means of solar radiation by appropriately selecting its surface finish.

Key: 1. Object; 2. Solar radiation; 3. Highly reflecting surface (impeding cooling by emission); 4. Dull black (causing heating by absorption).

Thus, the temperature of a transparent object or of one polished as smooth as a mirror will only be slightly elevated even during intense thermal radiation. The object is almost insensitive to thermal radiation, because in the first case, the rays are for the most part passing through the object and, in the latter case, the rays are reflected by the object, without being weakened or destroyed; i.e., without having lost some part of their energy. If, on the other hand, the surface of the object is dark and rough, it can neither pass the incident rays through nor reflect them: in this case, they must be absorbed and hence cause the body to heat up.

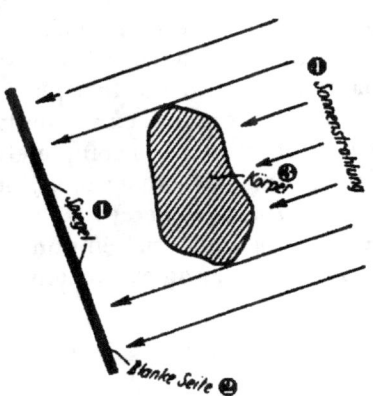

Figure 72. Heating of a body by protecting its shadow side against empty space by means of a mirror.

Key: 1. Mirror; 2. Polished side; 3. Object; 4. Solar radiation

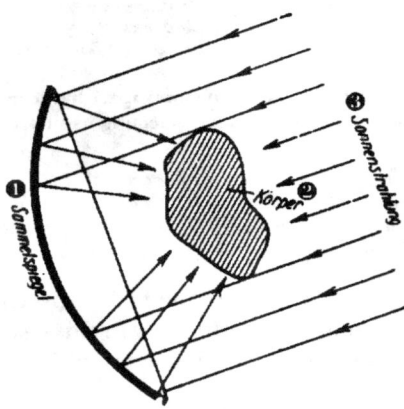

Figure 73. Intensive heating of an object by concentrating the rays of the sun on the object by means of a concave mirror.

Key: 1. Concave mirror; 2. Object; 3. Solar radiation.

This phenomenon is, however, not only valid for absorption but also for the release of heat through radiation: the brighter and smoother the surface of an object, the less is its ability to radiate and consequently the longer it retains its heat. On the other hand, with a dark, rough surface, an object can cool down very rapidly as a result of radiation.

The dullest black and least brightly reflecting surfaces show the strongest response to the various phenomena of thermal emission and absorption. This fact would make it possible to control the temperature of objects in empty space in a simple fashion and to a large degree.

If the temperature of an object is to be raised in space, then, as discussed above, its side facing the sun will be made dull black and the shadow side brightly reflecting (Figure 71); or the shadow side is protected against outer space by means of a mirror (Figure 72). If a concave mirror is used for this purpose, which directs the solar rays in an appropriate concentration onto the object, then the object's temperature could be increased significantly (Figure 73).

If, on the other hand, an object is to be cooled down in outer space, then its side facing the sun must be made reflective and its shadow side dull black (Figure 74); or it must be protected against the sun by means of a mirror (Figure 75). The object will lose more and more of its heat into space as a result of radiation because the heat can no longer be constantly replaced by conduction from the environment, as happens on Earth as a result of contact with the surrounding air, while replenishing its heat through incident radiation would be decreased to a minimum as a result of the indicated screening. In this manner, it should be possible to cool down an object to nearly absolute zero (-273° Celsius). This temperature could not be reached completely, however, because a certain amount of heat is radiated by fixed stars to the object on the shadow side; also, the mirrors could not completely protect against the sun.

By using the described radiation phenomena, it would be possible on the space station not only to provide continually the normal heat necessary for life, but also to generate extremely high and low temperatures, and consequently also very significant temperature gradients.

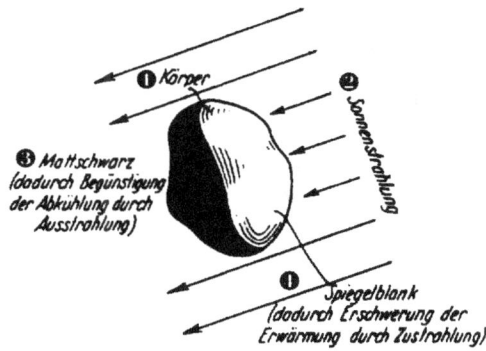

Figure 74. Cooling an object down in empty space by appropriately selecting its surface finish.

Key: 1. Object; 2. Solar radiation; 3. Dull black (promoting cooling through emission); 4. Highly reflective surface (impeding heating by radiation).

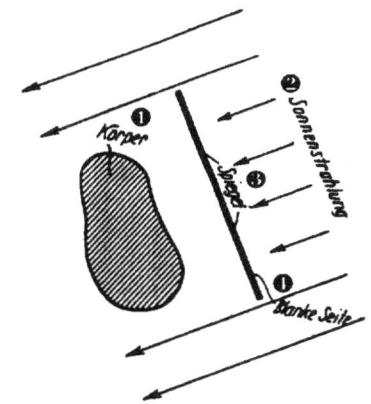

Figure 75. Cooling an object by protecting it against solar radiation by means of a mirror.

Key: 1. Object; 2. Solar radiation; 3. Mirror; 4. Reflective side.

Designing the Space Station

The physical conditions and potentials of empty space are now familiar to us. Here is an idea of how our space station would have to be designed and equipped: in order to simplify as far as possible the work to be performed in outer space when constructing this observatory (this work only being possible in space suits), the entire structure including its equipment would have to be assembled first on Earth and tested for reliability. Furthermore, it would have to be constructed in such a manner that it could easily be disassembled into its components and if at all possible into individual, completely furnished "cells" that could be transported into outer space by means of space ships and reassembled there without difficulty. As much as possible, only light-weight metals should be used as materials in order to lower the cost of carrying them into outer space.

The completed, ready-to-use structure would, in general, look as follows: primarily, it must be completely sealed and airtight against empty space, thus permitting internally normal atmospheric conditions to be maintained by artificial means. In order to reduce the danger of escaping air, which would happen if a leak occurred (e.g., as a result of an impacting meteor), the space station would be partitioned in an appropriate manner into compartments familiar from ship building.

Since all rooms are connected with one another and are filled with air, movement is easily possible throughout the inside of the space station. Space travelers can, however, only reach the outside into empty space by means of so-called air locks. This equipment, (used in caissons, diving bells, etc.) familiar from underwater construction, consists primarily

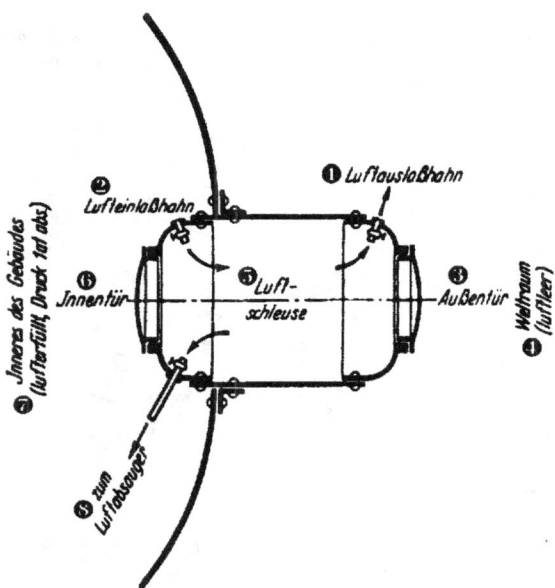

Figure 76. Basic layout of an air lock for moving from an air-filled room (e.g., the inside of the space station) to empty space. Drawing the air out of the lock during "outgoing" occurs mostly by pumping the air into the station for reasons of economy; only the residual air in the lock is exhausted into empty space. [Editor's Note: This second sentence was a footnote].

Key: 1. Air outlet valve; 2. Air intake valve; 3. Outside door; 4. Outer space (airless); 5. Air lock; 6. Inside door; 7. Inside of the station (air-filled, pressure of 1 at absolute); 8. To air pump.

of a small chamber that has two doors sealed airtight, one of which leads to the inside of the station and the other to the outside (Figure 76).

If for example, a space traveler wants to leave the space station ("outgoing" or egress), then, dressed in the space suit, he enters the lock through the inside door, the outside door being locked. Now the inside door is locked and the air in the lock is pumped out and finally exhausted, thus allowing the traveler to open the outer door and float out into the open. In order to reach the inside of the space station ("incoming" or ingress), the reverse procedure would have to be followed.

For operations and the necessary facilities of the space station, it is important to remember that absolutely nothing is available locally other than the radiation of the stars, primarily those of the sun; its rays, however, are available almost all the time and in unlimited quantity. Other substances particularly necessary for life, such as air and water, must be continually supplied from the Earth. This fact immediately leads to the following principle for the operation of the space station: exercise extreme thrift with all consumables, making abundant use instead of the energy available locally in substantial quantities in the sun's radiation for operating technical systems of all types, in particular those making it possible to recycle the spent consumables.

The Solar Power Plant

The solar power plant for that purpose (Figure 77) represents, therefore, one of the most important systems of the space station. It delivers direct current, is equipped with a storage battery and is comparable in principle to a standard steam turbine power plant of the same type. There are differences, however: the steam generator is now heated by solar radiation, which is concentrated by a concave mirror in order to achieve sufficiently high temperatures (Figure 77,

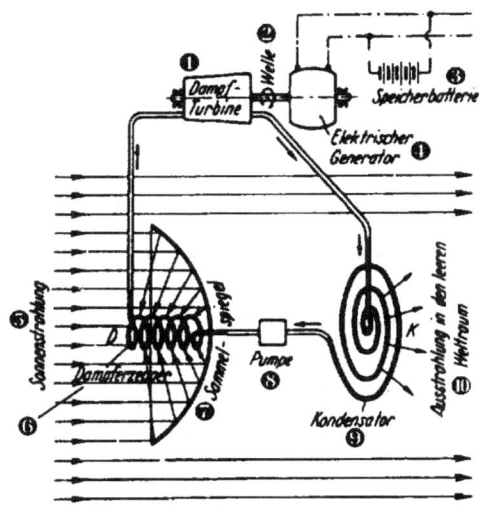

Figure 77. Diagram of the solar power plant of the space station.

Key: 1. Steam turbine; 2. Shaft; 3. Storage battery; 4. Electrical generator; 5. Solar radiation; 6. Steam generator; 7. Concave mirror; 8. Pump; 9. Condenser; 10. Radiation out into empty space.

D); and cooling of the condenser occurs only by radiating into empty space, so it must be opened towards empty space and shielded from the sun (Figure 77, K).

In accordance with our previous explanations, this causes both the steam generator and condenser to be painted dull black on the outside. In essence, both consist only of long, continuously curved metal pipes, so that the internal pipe walls, even in a weightless state, are always in sufficiently strong contact with the working fluid flowing through them (see Figure 77).

This working fluid is in a constant, loss-free circulation. Deviating from the usual practice, rather than water (steam), a highly volatile medium, nitrogen, is used in this case as a working fluid. Nitrogen allows the temperature of condensation to be maintained so low that the exceptional cooling potential of empty space can be used. Furthermore, an accidental escape of nitrogen into the rooms of the space station will not pollute its very valuable air.

Since it is only the size of the concave mirror that determines how much energy is being extracted from solar radiation, an appropriately efficient design of the power plant can easily ensure that sufficient amounts of electrical and also of mechanical energy are always available on the space station. Furthermore, since heat, even in great amounts, can be obtained directly from solar radiation, and since refrigeration, even down to the lowest temperatures, can be generated simply by radiating into empty space, the conditions exist to permit operation of all types of engineering systems.

Supplying Light

Lighting the space station can be accomplished very easily because this requires almost no mechanical equipment but can be achieved directly from the sun, which after all shines incessantly, disregarding pos-

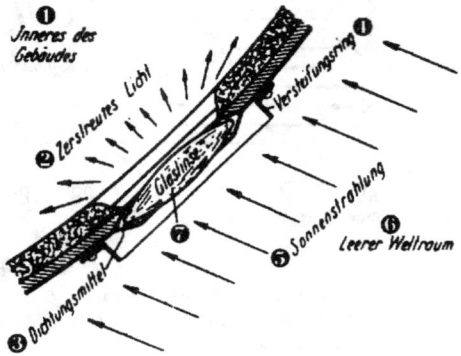

Figure 78. Lighting hatch.

Key: 1. Interior of the structure; 2. Dispersed light; 3. Sealing material; 4. Bracing; 5. Solar radiation; 6. Empty space; 7. Glass lens.

sible, yet in every case only short, passes by the space station through the Earth's shadow.

For this purpose, the walls have round openings similar to ship's bull's-eyes (Figures 60 and 61, L) with strong, lens-type glass windows (Figure 78). A milk-white coloring or frosting of the windows, and also an appropriate selection of the type of glass, ensure that sunlight is freed of all damaging radiation components, filtered in the same way as through the atmosphere, and then enters into the space station in a diffuse state. Therefore, the station is illuminated by normal daylight. Several of the bull's-eyes are equipped with special mirrors through which the sun's rays can be directed on the bull's-eyes when needed (Figure 79). In addition, artificial (electrical) lighting is provided by extracting current from the solar power plant.

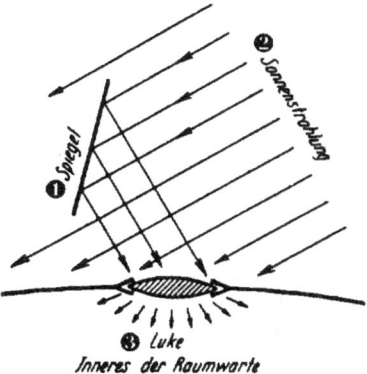

Figure 79. The mirror reflects the rays of the sun directly on the window.

Key: 1. Mirror; 2. Solar radiation; 3. Window and interior of the space station.

Supplying Air and Heat

Even heating the space station takes place by directly using solar radiation, more specifically, according to the principle of heating air simultaneously with ventilation.

For this purpose, the entire air of the space station is continuously circulating among the rooms requiring it and through a ventilation system where it is cleaned, regenerated, and heated. A large, electrically driven ventilator maintains air movement. Pipelines necessary for this process are also available. They discharge through small screened openings (Figures 60 and 61, O) into the individual rooms where the air is consumed. The ventilation system (Figure 80) is equipped similarly to the air renewing device suggested by Oberth. At first, the air flows through a dust filter. Then it arrives in a pipe cooled by radiating into outer space; the temperature in this pipe is lowered gradually to below -78° Celsius, thus precipitating the gaseous admixtures; more specifically, first the water vapor and later the carbon dioxide. Then, the air flows through a pipe heated by the concentrated rays of the sun, thus bringing it to the temperature necessary to keep the rooms warm. Finally, its oxygen and moisture contents are also brought to the proper levels, whereupon it flows back into the rooms of the space station.

This process ensures that only the oxygen consumed by breathing must be replaced and consequently resupplied from the ground; the non-consumed components of the air (in particular, its entire nitrogen portion) remain continually in use. Since the external walls of the space station do not participate in this heating procedure, these walls must be inhibited as much as possible from dissipating heat into outer space through radiation; for this reason, the entire station is highly polished on the outside.

Water Supply

The available water supply must also be handled just as economically: all the used water is collected and again made reusable through purification. For this purpose, large distillation systems are used in which the evaporation and subsequent condensation of the water is accomplished in a similar fashion as was previously described for the solar power plant: in pipes heated by the concentrated rays of the sun (Figure 77, D) and cooled by radiating into outer space (Figure 77, K).

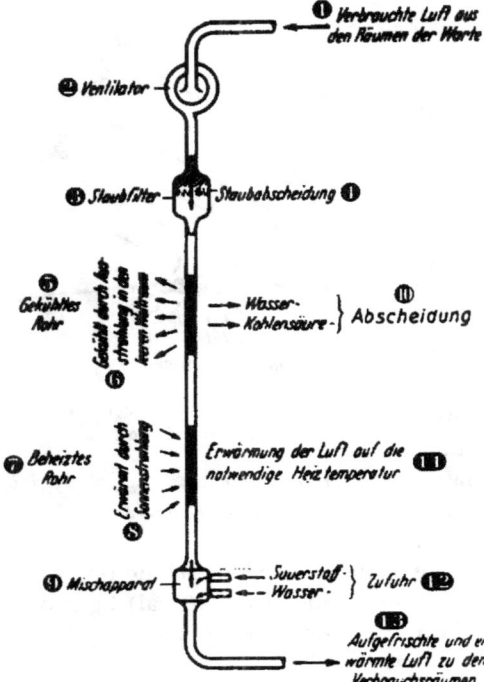

Figure 80. Schematic representation of a ventilation system. The cooled and heated pipes could be built, for example, similar to the ones shown in Figure 75, D and/or K.

Key: 1. Used air from the rooms of the space station; 2. Fan; 3. Dust filter; 4. Dust separation; 5. Cooled pipe; 6. Cooled by radiating into empty space; 7. Heated pipe; 8. Heated by solar radiation; 9. Mixer; 10. Water and carbon dioxide separation; 11. Heating the air to the required temperature; 12. Oxygen and water supply; 13. Regenerated and heated air to the rooms where it is consumed.

Long Distance Communications

The equipment for long distance communication is very important. Communication takes place either through heliograph signaling using a flashing mirror, electrical lamps, spot lights, colored disks, etc., or it is accomplished electrically by radio or by wires within the confined areas of the space station.

In communicating with the ground, use of heliograph signaling has the disadvantage of being unreliable because it depends on the receiving station on the Earth being cloudless. Therefore, the space station has at its disposal large radio equipment that makes possible both telegraph and telephone communications with the ground at any time. Overcoming a relatively significant distance as well as the shielding effect exerted by the atmosphere on radio waves (Heaviside layer),* are successful here (after selecting an appropriate direction of radiation) by using shorter, directed waves and sufficiently high transmission power, because conditions for this transmission are favorable since electric energy can be generated in almost any quantities by means of the solar power plant and because the construction of any type of antenna presents no serious problems as a result of the existing weightlessness.

Means of Controlling the Space Station

Finally, special attitude control motors ("momentum wheels") and thrusters are planned that serve both to turn the space station in any direction and to influence its state of motion as necessary. On the one hand, this option must exist to be able to maintain the space station in the desired orientation relative to the Earth as well as in the required position relative to the direction of the rays of the sun. For this purpose, not only must all those impulses of motion (originating from outside of the system!) that are inevitably imparted to the space station again and again in the traffic with space ships be continually compensated for, but the

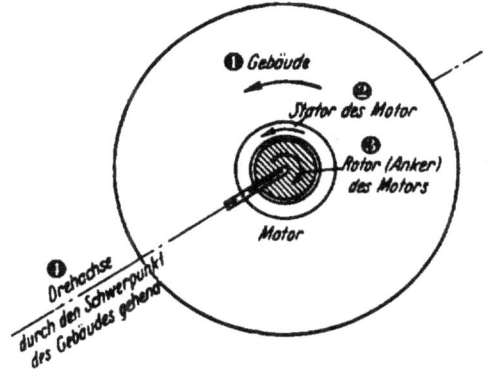

Figure 81. Operating characteristics of a thruster motor (see the text).

Key: 1. Station; 2. Stator of the motor; 3. Rotor (armature) of the motor; 4. Axis of rotation going through the center of mass of the station.

---

* An ionized layer in the upper atmosphere that reflects radio waves back to Earth. It was postulated by English physicist Oliver Heaviside (1850-1925) and U.S. electrical engineer Arthur E. Kennelly (1861-1939) in 1902. Also called the Kennelly-Heaviside layer and now referred to as the E region of the ionosphere.

effect of the Earth's movement around the sun must also be continually taken into account.

On the other hand, this is also necessary in order to enable the space station to satisfy its special tasks, which will be discussed later, because any changes of its position in space must be possible for performing many of these tasks and finally because the necessity can occasionally arise for repositioning the station in relation to the Earth's surface.

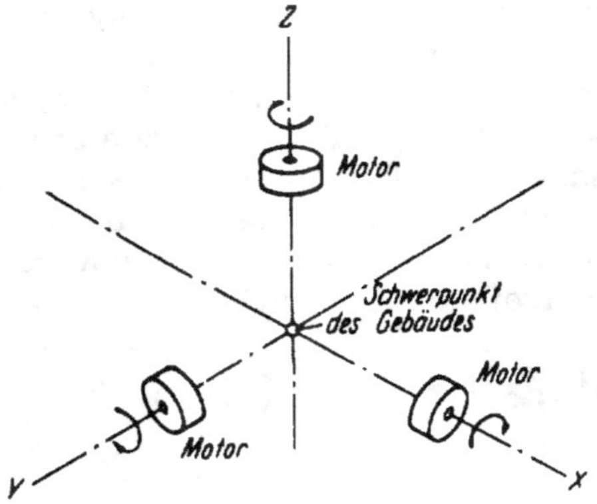

The attitude control motors are standard direct current electrical motors with a maximum rate of revolution as high as possible and a relatively large rotor mass. Special brakes make it possible rapidly to lower or shut off their operation at will. They are installed in such a manner that their extended theoretical axes of rotation go through the center of mass of the station.

Figure 82. Orientation of the attitude control motors (momentum wheels). The 3 axes are perpendicular to one another and go through the center of mass of the station.

Key: 1. Center of mass of the station.

Now, if an attitude control motor of this type is started (Figure 81), then its stator (the normally stationary part of the electrical motor), and consequently the entire station firmly connected to the motor rotate simultaneously with its rotor (armature) around the axis of the motor—however, in the opposite direction and, corresponding to the larger mass, much more slowly than the rotor. More specifically, the station rotates until the motor is again turned off; its rate of rotation varies depending on the rate of revolution imparted to the rotor of the motor. (In the present case, there is a "free system," in which only internal forces are active.) Since these motors are oriented in such a fashion that their axes are perpendicular to one another like those of a right angle, three dimensional coordinate system (Figure 82), the station can be rotated in any arbitrary direction due to their cooperative combined effects.

The thrusters are similar both in construction and in operating characteristics to the propulsion systems of the space ships described previously.[26] They are, however, much less powerful than those described, cor-

---

26. See pages 33-34.

responding to the lesser demands imposed on them (the accelerations caused by them need not be large). They are positioned in such a manner that they can impart an acceleration to the station in any direction.

Partitioning the Space Station into Three Entities

It would, no doubt, be conceivable to design technical equipment that makes possible staying in empty space despite the absence of all materials; however, even the absence of gravity would (at least in a physical sense, probably otherwise also) not present any critical obstacle to the sustenance of life, if the various peculiarities resulting from space conditions are taken into consideration in the manner previously indicated.

However, since the weightless state would be associated in any case with considerable inconveniences and could even perhaps prove to be dangerous to health over very long periods of time, artificial replacement of gravity is provided for in the space station. In accordance with our previous discussion, the force of gravity, being an inertial force, can only be influenced, offset or replaced by an inertial force, more specifically, by centrifugal force if a permanent (stabile) state is to result. This very force allows us to maintain the space station in its vertiginous altitude, so to speak, and to support it there. However, since the latter also leads at the same time to complete compensation of the gravitational state in the space station itself, the centrifugal force now is used again (however, in a different manner) to replace the missing gravitational state.

Basically, this is very easy to accomplish: only those parts of the station in which the centrifugal force and consequently a gravitational state are to be produced must be rotated at the proper speed around their center of mass (center of gravity). At the same time, it is more difficult to satisfy the following requirement: the space traveller must be able to exit and enter the station, connect cables and attach large concave mirrors simply and safely when some parts of the station are rotating. Another requirement is that it be possible also to reposition the entire station not only relative to the sun's rays, but also according to the demands of remote observations.

These conditions lead to a partitioning of the space station into three individual entities: first, the "habitat wheel," in which a man-made gravitational state is continually maintained through rotation, thus offering the same living conditions as exist on Earth; it is used for relaxing and for the normal life functions; second, the "observatory"; and third, the "machine room." While retaining the weightless state, the latter two are only equipped in accordance with their special functions; they provide the personnel on duty with a place for performing their work, but only for a short stay.

However, this partitioning of the space station makes it necessary to apply special procedures in order to compensate for the mutual attraction of the individual objects, because even though this is very slight due to the relative smallness of the attracting masses, the mutual attraction would nevertheless lead to a noticeable approach over a longer period (perhaps in weeks or months) and finally even to the mutual impact of individual objects of the space station. The individual objects, therefore, must either:

be positioned as far as possible from one another (at several hundred or thousand meters distance), so that the force of mutual attraction is sufficiently low; from time to time the approach that is occurring nonetheless can be compensated for by means of thrusters, or;

be as close as possible to one another and be mutually braced in a suitable manner to keep them separated. In this study, we decided on the first alternative (Figure 94).

The Habitat Wheel

As is generally known, both the velocity of rotation and the centrifugal force at the various points of a rotating object are proportional to the distance from its center of rotation, the axis (Figure 83); i.e., the velocity is that much greater, the further the point in question is distant from the axis and that much less, the closer it is to the axis; it is equal to zero on the theoretical axis of rotation itself.

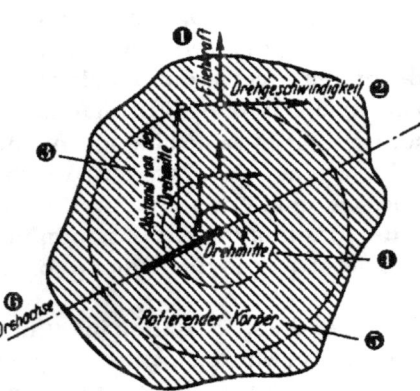

Figure 83. Velocity of rotation and centrifugal force on a rotating object.

Key: 1. Centrifugal force; 2. Velocity of rotation; 3. Distance from the center of rotation; 4. Center of rotation; 5. Rotating object; 6. Axis of rotation.

Accordingly, the rotating part of the space station must be structured in such a manner that its air lock and the cable connections in the center of the entire structure are in the axis of rotation because the least motion exists at that point, and that those parts, in which a gravitational effect is to be produced by centrifugal force, are distant from the axis on the perimeter because the centrifugal force is the strongest at that point.

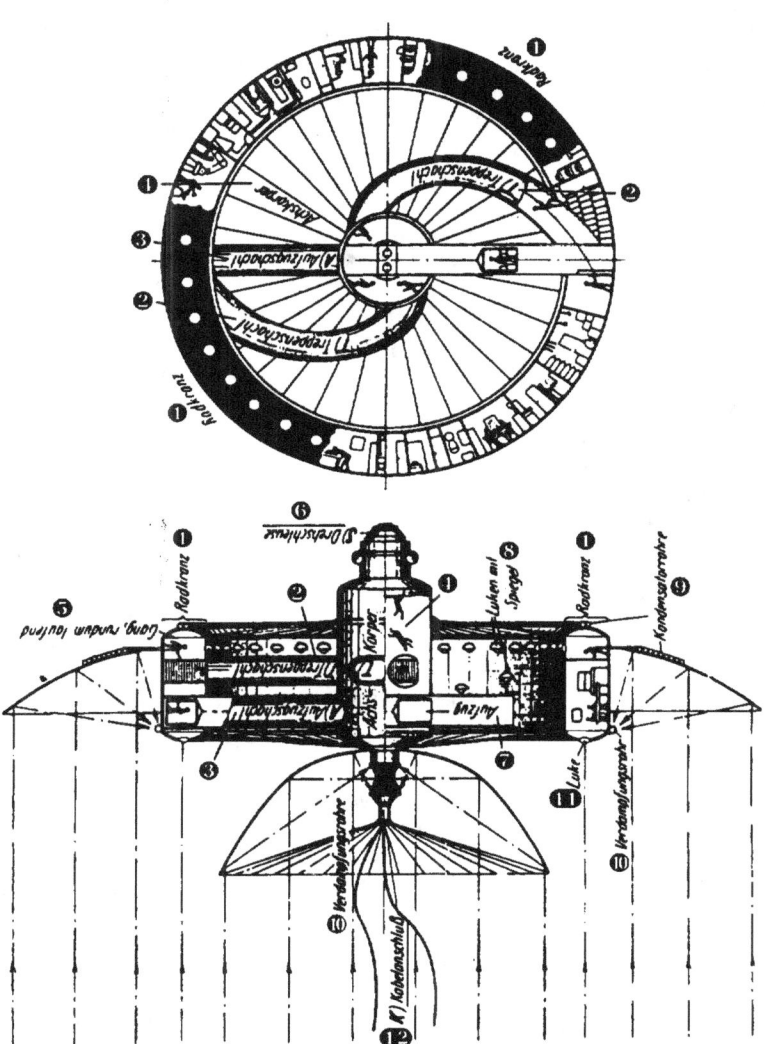

Figure 84. The Habitat Wheel. Left: axial cross section. Right: view of the side constantly facing the sun, without a concave mirror, partially in cross section.

Key: 1. Wheel rim; 2. Well of the staircase; 3. Elevator shaft; 4. Axial segment; 5. Circular corridor; 6. Turnable air lock; 7. Elevator; 8. Bull's-eyes with mirrors; 9. Condenser pipes; 10. Evaporation tube; 11. Bull's-eye (window); 12. Cable connection.

These conditions are, however, best fulfilled when the station is laid out in the shape of a large wheel as previously indicated (Figures 84, 89 and 90): the rim of the wheel is composed of cells and has the form of a ring braced by wire spokes towards the axis. Its interior is separated into individual rooms by partitions; all rooms are accessible from a wide corridor going around the entire station. There are individual rooms, larger sleeping bays, work and study areas, a mess hall, laboratory, workshop, dark room, etc., as well as the usual utility areas, such as a kitchen, bath room, laundry room and similar areas. All rooms are furnished with modern day comforts; even cold and warm water lines are available. In general, the rooms are similar to those of a modern ship. They can be furnished just like on Earth because an almost normal, terrestrial gravitational state exists in these rooms.

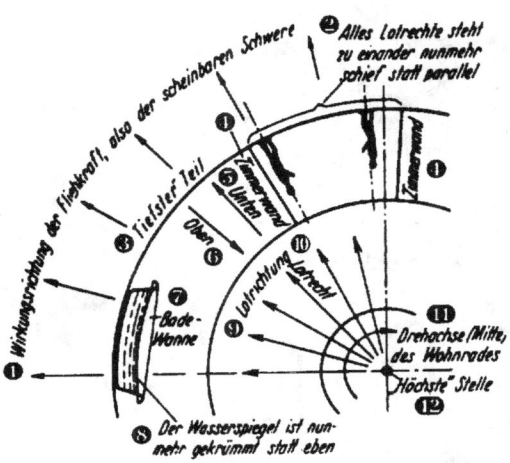

Figure 85. Directional relationships in the habitat wheel.

Key: 1. Direction of the centrifugal force, that is, of apparent gravity; 2. Everything vertical is tilted instead of parallel; 3. "Lowest" region; 4. Partition; 5. Down; 6. Up; 7. Bathtub; 8. The water level is curved instead of straight (flat); 9. Vertical direction; 10. Vertical; 11. Axis of rotation (center) of the habitat wheel; 12. "Highest" point.

However, to create this gravitational state, the entire station, assuming a diameter of 30 meters, for example, must rotate in such a manner that it performs a complete rotation in about 8 seconds, thus producing a centrifugal force in the rim of the wheel that is just as large as the gravitational force on the Earth's surface.

While the force of gravity is directed towards the center of mass, the centrifugal force, on the other hand, is directed away from the center. Therefore, "vertical" in the habitat wheel means the reverse of on Earth: the radial direction from the center (from the axis of rotation) directed outward (Figure 85). Accordingly, "down" now points towards the perimeter and at the same time to the "lowest" part, while "up" now points towards the axis and at the same time to the "highest" point of this man-made celestial body. Taking its smallness into account, the radial orientation of the vertical direction, which in most cases is irrelevant on the

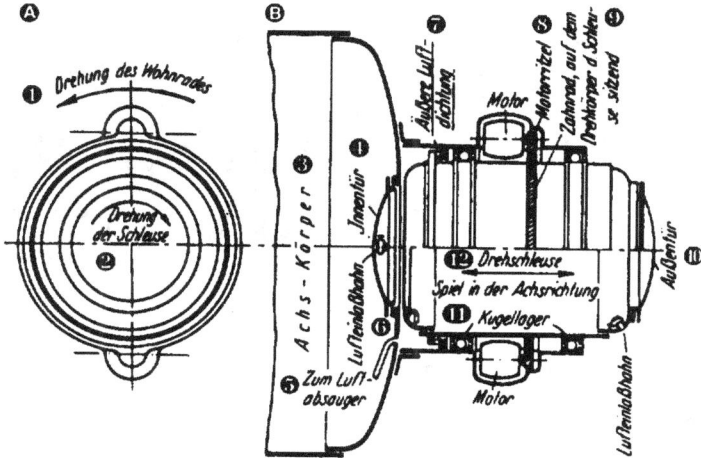

Figure 86. a) Top view onto the external door of the rotating air lock of the habitat wheel. b) Axial cross section through the rotating air lock of the habitat wheel.
(See Figure 84 and the text.) The ball bearings are designed in such a manner that they allow movement in the direction of the axis through which closing and/or releasing is possible of the external air seal which connects the air lock airtight to the inside of the habitat wheel when the inside door is open.

Key: 1. Rotation of the habitat wheel; 2. Rotation of the air lock; 3. Axial segment; 4. Inside door; 5. To the air pump; 6. Air intake valve; 7. External air seal; 8. Motor pinion gear; 9. Gear on the rotor of the lock; 10. Outside door; 11. Ball bearing; 12. Rotating air lock, movement in the axial direction.

Earth due to its size, now clearly becomes evident in the space station. The consequence of this is that all "vertical" directions (such as those for human beings standing erect, the partitions of the rooms, etc.) are now convergent instead of parallel to one another, and everything "horizontal" (e.g., water surface of the bathtub) appears curved instead of flat (see Figure 85).

A further peculiarity is the fact that both the velocity of rotation and the centrifugal force, as a result of their decrease towards the center of rotation, are somewhat less at the head of a person standing in the habitat wheel than at his feet (by approximately $1/9$ for a wheel diameter of 30 meters) (Figure 83). The difference in the centrifugal forces should hardly be noticeable, while that of the velocities of rotation should be noticeable to some degree, especially when performing up and down (i.e., radial) movements, such as lifting a hand, sitting down, etc.

However, all of these phenomena make themselves felt that much less, the larger the diameter of the wheel. In the previously selected case (30 meters in diameter), only a slight effect would be perceptible.

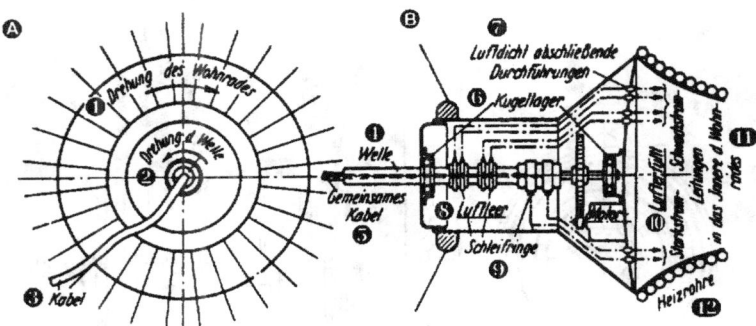

Figure 87. A. Top view onto the cable connection of the habitat wheel. B. Axial cross section through the cable connection of the habitat wheel.
(See Figures 84, K, and the text.)

Key: 1. Rotation of the habitat wheel; 2. Rotation of the shaft; 3. Cable; 4. Shaft; 5. Compound cable; 6. Ball bearings; 7. Passageways sealed airtight; 8. Vacuum; 9. Sliding contact rings; 10. Pressurized; 11. High and low voltage lines on the inside of the habitat wheel; 12. Heating tube.

the air lock. Then, the latter begins to rotate by electrical power—opposite to the direction of rotation of the habitat wheel—until it reaches the same rotational speed as the habitat wheel. As a result, the air lock is stationary in relation to outer space and can now be departed just as if the habitat wheel were not even rotating. The process is reversed for "incoming." With some training, rotating the air lock can, however, be dispensed with because the habitat wheel rotates only relatively slowly at any rate (one complete revolution in approximately 8 seconds in the previously assumed case with a 30 meter diameter of the wheel).

Even the cable connection at the other side of the axle segment is designed in a basically similar manner in order to prevent the cable from becoming twisted by the rotation of the habitat wheel. For this reason, the cable extends out from the end of a shaft (Figure 87), which is positioned on the theoretical axis of rotation of the habitat wheel and is continually driven by an electrical motor in such a manner that it rotates at exactly the same speed as the habitat wheel—but in the opposite direction. As a result, the shaft is continually stationary in relation to outer space. The cable extending from the shaft cannot, in fact, be affected by the rotation of the habitat wheel.

Stairs and electrical elevators installed in tubular shafts connect the axial segment and the rim of the wheel. These shafts run "vertically" for the elevators, i.e., radially (Figure 84, A). On the other hand, for the stairs, which must be inclined, the shafts are—taking the divergence of the vertical direction into account—curved along logarithmic spirals that

tioned on the theoretical axis of rotation of the habitat wheel and is continually driven by an electrical motor in such a manner that it rotates at exactly the same speed as the habitat wheel—but in the opposite direction. As a result, the shaft is continually stationary in relation to outer space. The cable extending from the shaft cannot, in fact, be affected by the rotation of the habitat wheel.

Stairs and electrical elevators installed in tubular shafts connect the axial segment and the rim of the wheel. These shafts run "vertically" for the elevators, i.e., radially (Figure 84, A). On the other hand, for the stairs, which must be inclined, the shafts are—taking the divergence of the vertical direction into account—curved along logarithmic spirals that gradually become steeper towards "up" (towards the center) (Figures 88 and 84, T) because the gravitational effect (centrifugal force) decreases more and more towards that point. By using the stairs and/or elevators in an appropriately slow manner, the transition can be performed gradually and arbitrarily between the gravitational state existing in the rim of the wheel and the absence of gravity in outer space.

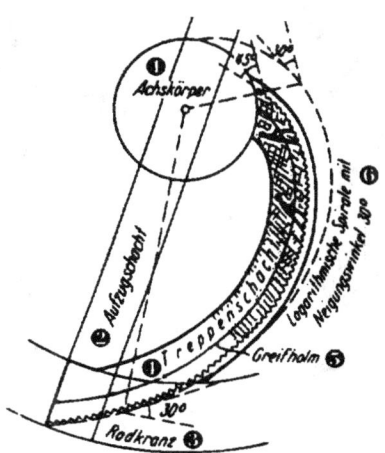

Figure 88. Well of the habitat wheel staircase.

Key: 1. Axial beam; 2. Elevator shaft; 3. Rim of the wheel; 4. Well of the staircase; 5. Railing; 6.

Supplying the habitat wheel with light, heat, air and water takes place in the fashion previously specified in general for the space station by employing the engineering equipment described there. The only difference being that the wall of the wheel rim always facing the sun also acts to heat the habitat wheel;[27] for this reason, this wall is colored dull black (Figures 89 and 84), in contrast to the otherwise completely highly polished external surfaces of the station. A small solar power plant sufficient for emergency needs of the habitat wheel is also available.

All storage rooms and tanks for adequate supplies of air, water, food and other materials, as well as all mechanical equipment are in the wheel rim. The concave mirrors associated with this equipment and the dull black colored steam generator and condenser pipes are attached to the

---

27. Naturally, the sun's help could also be dispensed with and supplying the heat of the habitat wheel could also be provided solely by means of air heating. The entire rim of the wheel would then have to be highly polished.

Figure 89. Total view of the side of the habitat wheel facing the sun. The center concave mirror could be done away with and replaced by appropriately enlarging the external mirror.

habitat wheel on the outside in an appropriate manner and are rotating with the habitat wheel (Figures 84, 89 and 90).

Finally, attitude control motors and thrusters are provided; besides the purposes previously indicated, they will also generate the rotational motion of the habitat wheel and stop it again; they can also control the rate of rotation.

The Observatory and Machine Room

The decisive idea for the habitat wheel—creating living conditions as comfortable as possible—must be of secondary importance for the ob-

Figure 90. Total view of the shadow side of the habitat wheel.

servatory and machine room compared to the requirement for making these systems primarily suitable for fulfilling their special tasks. For this reason, eliminating the weightless state is omitted, as noted previously, for these systems.

Primarily, it is important for the observatory (Figure 91) that any arbitrary orientation in space, which is necessitated by the observations to be carried out, can easily be assumed. It must, therefore, be completely independent of the sun's position; consequently, it may not have any of the previously described equipment that is powered by solar radiation. For this reason, ventilation and the simultaneous heating of the observatory as well as its electrical supply take place from the machine room;

Figure 91. An example of the design of an observatory. Taking into account the overpressure of 1 atm. existing inside, the observatory resembles a boiler. The air lock, two electrical cables (left), the flexible air tube (right) and the bull's-eyes can be seen.

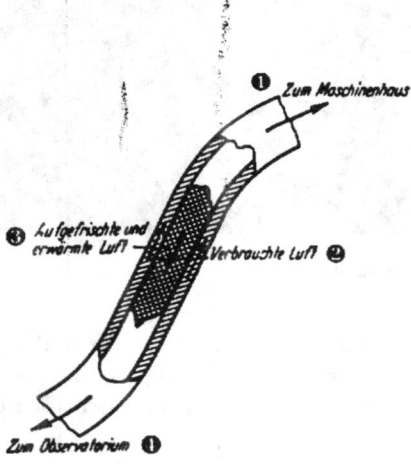

Figure 92. The flexible tube for connecting the observatory with the ventilation system in the machine room.

Key: 1. To the machine room; 2. Spent air; 3. Regenerated and heated air; 4. To the observatory.

consequently, both units are connected also by a flexible tube as well as a cable (Figures 91 and 92). Nevertheless, a precaution is taken to ensure that the ventilation of the observatory can also be carried out automatically in an emergency by employing purification cartridges, as is customary in modern diving suits.

The observatory contains the following equipment: primarily, remote observation equipment in accordance with the intended purpose of this unit and, furthermore, all controls necessary for remote observations, like those needed for the space mirror (see the following). Finally, a laboratory for performing experiments in the weightless state is also lo-

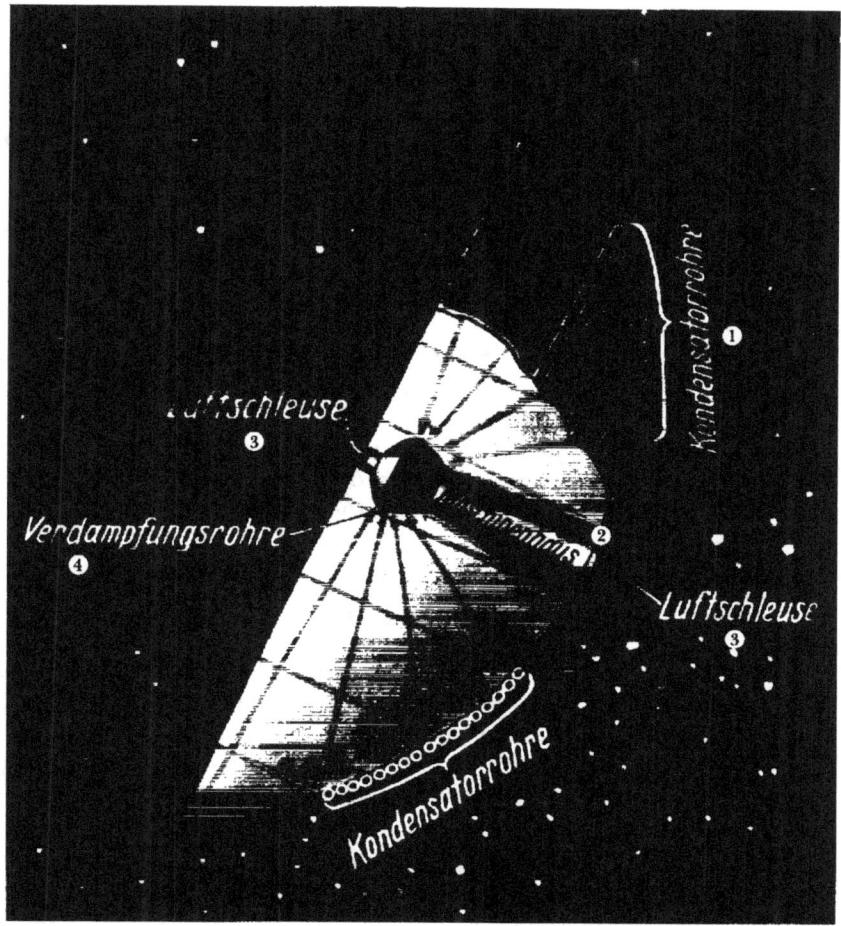

Figure 93. Example of the design of the machine room shown in the axial cross section.

Key: 1. Condenser pipes; 2. Machine room; 3. Air lock; 4. Evaporating pipes.

cated in the observatory.

The machine room is designed for housing the major mechanical and electrical systems common to the entire space station, in particular those that serve for the large-scale utilization of the sun's radiation. Primarily, it contains the main solar power plant including storage batteries. Furthermore, all of the equipment in the large transmission station is located here, and finally, there is a ventilation system, which simultaneously supplies the observatory.

Collecting solar energy takes place through a huge concave mirror firmly connected to the machine room (Figure 93), in whose focal point the evaporating and heating pipes are located, while the condenser and cooling pipes are attached to its back side. The orientation of the machine room is, therefore, determined beforehand: the concave mirror must always squarely face the sun.

Lighting of both the observatory and machine room is achieved in the manner already described in general for the space station. All external surfaces of the units are highly polished in order to reduce the cooling effect. Finally, both units are also equipped with attitude control motors and thrusters.

Kitchens, water purification systems, washing facilities, and similar systems are missing, however, because of the very troublesome properties of liquids in the weightless state. The habitat wheel is available for eating and personal hygiene. Necessary food and beverages in the observatory and machine room must be brought in from the habitat wheel, prepared in a manner compatible with the weightless state.

Providing for Long-Distance Communications and Safety

Communicating among the individual components of the space station takes place in the manner previously indicated either through signaling with lights or by either radio or wire. Accordingly, all three substations are equipped with their own local radio stations and, furthermore, are connected to one another by cables that include electric power lines.

Finally, each one of the three substations carries reserve supplies of food, oxygen, water, heating material and electricity (stored in spare batteries) in such a manner that it can house the entire crew of the space station for some time in an emergency, if, for instance, each of the other two substations should become unusable at the same time through an accident. In this manner, the tri-partitioning of the space station, originally chosen for technical reasons, also contributes considerably to safety. In order to enhance the latter still further, provisions are made to ensure that each substation not only can communicate with the ground through the central radio station, but also independently via its own flashing mirror system.

Partitioning the Space Station into Two Entities

Instead of three parts, the space station could also be partitioned into only two entities by combining the habitat wheel and the machine room.

Basically, this would be possible because the orientation in outer space for these two units is determined only by the direction of the sun's rays; more specifically, it is determined in the same manner.

If the mirror of the machine room is to be exempted from participating in the (for its size) relatively rapid rotation of the habitat wheel, then, for example, the habitat wheel and machine room (including its mirror) could both be rotated around a common axis of rotation—but in a reverse sense. Or the habitat wheel and machine room could be completely integrated into one structure, and the large mirror of the machine room alone could be rotated around its axis of rotation, also in an opposite direction. Other methods could also be employed.

The advantages of a two-component space station would be as follows:

1. Movement within the space station is simplified.

2. The provisions necessary in a separated partitioning to compensate for the mutual attraction between habitat wheel and machine room are no longer needed.

3. The rotational motion of the habitat wheel can now be produced, changed and stopped through motor power instead of thrusters—without any expenditure of propellants—because now the entire machine room and/or its large mirror are available as a "counter mass" for this purpose (consequently, the reverse rotational direction of the mirror).

These advantages are countered by the disadvantage that significant design problems result, but these are solvable. We want to refrain from examining any further this partitioning of the space station in more detail here in order not to complicate the picture obtained of it up to this point.

The Space Suit

Both for assembly and operation of the space station (moving between individual entities and to and from the space ships, performing varying tasks, etc.) it is necessary to be able to remain outside of the enclosed rooms in open space. Since this is only possible using the previously mentioned space suits, we have to address these suits in more detail.

As previously explained, they are similar to the modern diving and/or gas protective suits. But in contrast to these two suits, the space suit garment must not only be airtight, resistant to external influences and built in such a manner that it allows movement to be as unrestricted as possible; additionally, it must have a large tensile strength because a gas pressure (overpressure of the air in relation to empty space) of one full atmosphere exists within the garment. And moreover, it should be insensitive to the extremely low temperatures that will prevail in empty space due to heat loss by thermal emission. The garment must neither become brittle nor otherwise lose strength. Without a doubt, very significant requirements will be imposed on the material of such a space suit.

In any case, the most difficult problem is the protection against cold; or, more correctly stated, the task of keeping the loss of heat through radiation within acceptable limits. One must attempt to restrict the capability of the garment to radiate to a minimum. The best way of attaining this goal would be to give the suit in its entirety a high polish on the outside. It would then have to be made either completely of metal or at least be coated with a metal. However, an appropriately prepared flexible material insensitive to very low temperatures would perhaps suffice as a garment, if it is colored bright white on the outside and is as smooth as possible.

Nevertheless, the advantage of a material of this nature may not be all that great as far as the freedom of movement is concerned, because even when the garment used is flexible, it would be stiff—since the suit is inflated (taut) as a result of the internal overpressure—such that special precautions would have to be taken to allow sufficient movement, just as if the garment were made of a solid material, such as metal. The all-metal construction would appear to be the most favorable because much experience from the modern armor diving suits is available regarding the method of designing such stiff suits; furthermore a structure similar to flexible metal tubes could possibly also be considered for space use.

We will, therefore, assume that the space suits are designed in this manner. As a result of a highly polished external surface, their cooling due to thermal emission is prevented as much as possible. Additionally, a special lining of the entire suit provides for extensive thermal insulation. In case cooling is felt during a long stay on the outside, it is counteracted through inradiation from mirrors on the shadow side of the space suit. Supplying air follows procedures used for modern deep sea divers. The necessary oxygen bottles and air purification cartridges are carried in a metal backpack.

Since voice communication through airless space is possible only via telephones and since a connection by wires would be impractical for this purpose, the space suits are equipped with radio communication gear: a small device functioning as sender and receiver and powered by storage batteries is also carried in the backpack for this purpose. The microphone and the head phone are mounted firmly in the helmet. A suitably installed wire or the metal of the suit serves as an antenna. Since each individual unit of the space station is equipped for local radio communication, spacefarers outside the station can, therefore, speak with each other as well as with the interior of any of the space station units, just like in the air-filled space—however, not by means of air waves, but through ether waves.

For special safety against the previously described danger of "floating away into outer space" threatened during a stay in the open, the local radio stations are also equipped with very sensitive alarm devices that respond, even at great distances, to a possible call for help from inside a space suit.

In order to prevent mutual interference, various wavelengths are allocated to the individual types of local radio communications; these wavelengths can be tuned in easily by the radio devices in the space suits. Small hand-held thrusters make possible random movements. Their propellant tanks are also located in the backpack along with the previously described devices.

The Trip to the Space Station

The traffic between the Earth and the space station takes place through rocket-powered space ships, like those described in general in the first part of this book. It may complete the picture to envision such a trip at least in broad outlines:

The space ship is readied on the Earth. We enter the command room, a small chamber in the interior of the fuselage where the pilot and passengers stay. The door is locked airtight from the inside. We must lie down in hammocks. Several control actions by the pilot, a slight tremor in the vehicle and in the next moment we feel as heavy as lead, almost painfully the cords of the hammocks are pressed against the body, breathing is labored, lifting an arm is a test of strength: the ascent has begun. The propulsion system is working, lifting us up at an acceleration of 30 m/sec^2, and causing us to feel an increase of our weight to four times its normal value. It would have been impossible to remain standing upright under this load.

It does not take long before the feeling of increased gravity stops for a moment, only to start again immediately. The pilot explains that he has just jettisoned the first rocket stage, which is now spent, and started the second stage. Soon, new controlling actions follow: as explained by the pilot, we have already attained the necessary highest climbing velocity; for this reason, the vehicle was rotated by 90°, allowing the propulsion system to act now in a horizontal direction in order to accelerate us to the necessary orbital velocity.

Very soon, we have attained this velocity. Only some minutes have elapsed since launch; however, it seems endless to us, [given] that we had to put up with the strenuous state of elevated gravity. The pressure on us is gradually diminishing. First we feel a pleasant relief; then, however, an oppressive fear: we believe we are falling, crashing into the depths. The brave pilot attempts to calm us: he has slowly turned the propulsion system off. Our motion takes place now only by virtue of our own kinetic force, and what was sensed as free fall is nothing other than the feeling of weightlessness, something that we must get used to whether we like it or not. Easier said than done; but since we have no other choice, we finally succeed.

In the meantime, the pilot has acutely observed with his instruments and consulted his tables and travel curves. Several times the propulsion system was restarted for a short time: small orbital corrections had to be made. Now the destination is reached. We put on space suits, the air is vented from the command room, the door is opened. Ahead of us at some distance we see something strange, glittering in the pure sunlight like medallions, standing out starkly in the deep black, star-filled sky: the space station (Figure 94).

However, we have little time to marvel. Our pilot pushes away and floats toward the space station. We follow him, but not with very comfortable feelings: an abyss of almost 36,000 km gapes to the Earth! For the return trip, we note that our vehicle is equipped with wings. These were carried on board in a detached condition during the ascent and have now been attached, a job presenting no difficulties due to the existing weightlessness.

We re-enter the command room of the space ship; the door is closed; the room is pressurized. At first the propulsion system begins to work at a very low thrust: a slight feeling of gravity becomes noticeable. We must again lie down in the hammocks. Then, little by little more thrusters are switched on by the pilot, causing the sensation of gravity to increase to higher and higher levels. We feel it this time to be even heavier than before, after we had been unaccustomed to gravity over a longer period. The propulsion system now works at full force, and in a horizontal but opposite direction from before; our orbital velocity and consequently the centrifugal force, which had sustained us during the stay in the space station, must be decreased significantly to such a degree that we are freely falling in an elliptical orbit towards the Earth. A weightless state exists again during this part of the return trip.

In the meantime, we have come considerably closer to the Earth. Gradually, we are now entering into its atmosphere. Already, the air drag makes itself felt. The most difficult part of this trip is beginning: the landing. Now by means of air drag, we have to brake our travel velocity—which has risen during our fall to Earth up to around 12 times the velocity of a projectile—so gradually that no overheating occurs during the landing as a result of atmospheric friction.

Figure 94. The complete space station with its three units, seen through the door of a space ship. In the background—35,900 km distant—is the Earth. The center of its circumferential circle is that point of the Earth's surface on the equator over which the space station continually hovers (see pages 73-74). As assumed in this case, the space station is on the meridian of Berlin, approximately above the southern tip of Cameroon.

As a precautionary measure, we have all buckled up. The pilot is very busy controlling the wings and parachutes, continuously determining the position of the vehicle, measuring the air pressure and outside temperature, and performing other activities. For several hours, we orbit our planet at breakneck speed: in the beginning, it is a head-down flight at an altitude of approximately 75 km; later, with a continual decrease of the velocity, we approach the Earth more and more in a long spiral and, as a result, arrive in deeper, denser layers of air; gradually, the terrestrial feeling of gravity appears again, and our flight transitions into a normal gliding flight. As in a breakneck race, the Earth's surface rushes by underneath: in only half hours, entire oceans are crossed, continents traversed. Nevertheless, the flight becomes slower and slower and we come closer to the ground, finally splashing down into the sea near a harbor.

Special Physical Experiments

And now to the important question: What benefits could the described space station bring mankind! Oberth has specified all kinds of interesting proposals in this regard and they are referenced repeatedly in the following. For example, special physical and chemical experiments could be conducted that need large, completely airless spaces or require the absence of gravity and, for that reason, cannot be performed under terrestrial conditions. Furthermore, it would be possible to generate extremely low temperatures not only in a simpler fashion than on the Earth, but absolute zero could also be approached much more closely than has been possible in our refrigeration laboratories—to date, approximately 1° absolute, that is, -272° Celsius, has been attained there—because, besides the technique of helium liquefaction already in use for this purpose, the possibility of a very extensive cooling by radiating into empty space would be available on the space station.

The behavior of objects could be tested under the condition of an almost complete absence of heat, something that could lead to extremely valuable conclusions about the structure of matter as well as about the nature of electricity and heat, as the experiments of that type carried out previously in our refrigeration laboratories would lead us to expect. Probably even practical benefits—perhaps even to the grandest extent—would also result as a further consequence of these experiments. In this context, we could think of the problem, for example, of discovering a method for using the enormous amounts of energy bound up in matter.

Finally, in consideration of the special potentials offered by a space station, the problems of polar light, of cosmic rays, and of some other natural phenomena not yet fully explained could be brought to a final clarification.

Telescopes of Enormous Size

As has been previously explained, due to the absence of an atmosphere no optical barrier exists in empty space to prevent using telescopes of unlimited sizes. But also from the standpoint of construction, conditions are very favorable for such instruments due to the existing weightlessness. The electrical power necessary for remotely controlling the instruments and their components is also available in the space station.

Thus, for example, it would be possible to build even kilometer-long reflecting telescopes simply by positioning electrically adjustable, parabolic mirrors at proper distances from the observer in empty space. These and similar telescopes would be tremendously superior to the best ones available today on Earth. Without a doubt, it can be stated that almost no limits would exist at all for the performance of these instruments, and consequently for the possibilities of deep space observations.

Observing and Researching the Earth's Surface

Everything even down to the smallest detail on the Earth's surface could be detected from the space station using such powerful telescopes. Thus, we could receive optical signals sent from Earth by the simplest instruments and, as a result, keep research expeditions in communication with their home country, and also continually follow their activities. We could also scan unexplored lands, determining the make up of their soil, obtaining general information about their inhabitation and accessibility and, as a result, accomplish valuable preliminary work for planned research expeditions, even making available to these expeditions detailed photographic maps of the new lands to be explored.

This short description may show that cartography would be placed on an entirely new foundation, because by employing remote photography from the space station not only entire countries and even continents could be completely mapped in a simple fashion, but also detailed maps of any scale could be produced that would not be surpassed in accuracy even by the most conscientious work of surveyors and map makers. Land surveys of this type would otherwise take many years and require significant funding. The only task remaining then for map makers would be to insert the elevation data at a later date. Without much effort, very accurate maps could thus be obtained of all regions of the Earth still fairly unknown, such as the interior of Africa, Tibet, North Siberia, the polar regions, etc.

Furthermore, important marine routes—at least during the day and as far as permitted by the cloud cover—could be kept under surveillance in order to warn ships of impending dangers, such as floating icebergs, approaching storms and similar events, or to report ship accidents immediately. Since the movement of clouds on more than one-third of the entire Earth's surface could be surveyed at one time from the space station and at the same time cosmic observations not possible from the Earth could be performed, an entirely new basis should also result for weather forecasting.

And not the least important point is the strategic value of the possibilities of such remote observations: spread out like a war plan, the entire deployment and battle area would lie before the eyes of the observer in the space station! Even when avoiding every movement during the day as far as possible, the enemy would hardly be successful in hiding his intentions from such "Argus eyes."

Exploring the Stars

The most exciting prospects for remote observation from the space station exist for astronomy, because in this case, besides the possibility of using large telescopes at will, there are two other advantages: the radiations from the stars arrive completely unweakened and undistorted, and the sky appears totally black. Thus, for example, the latter condition would permit carrying out all those observations of the sun that can be performed on the Earth only during a total solar eclipse by simply occulting the solar disk using a round black screen.

Our entire solar system including all its planets, planetoids, comets, large and small moons, etc. could be studied down to the smallest detail. Even both ("inner") planets, Venus and Mercury, which are close to the sun could be observed just as well as the more distant ("outer") planets, observations that are not possible from the Earth due to dawn and dusk, a problem already mentioned. Therefore, the surfaces of at least all the near celestial bodies (Moon, Venus, Mars, Mercury), as far as they are visible to us, could be precisely studied and topographically mapped by remote photography. Even the question of whether the planets are populated, or at least whether they would be inhabitable, could probably be finally decided in this manner.

The most interesting discoveries would, however, presumably be made in the world of the fixed stars. Many unsolved puzzles at these extreme distances would be solved, and our knowledge of the functioning of the world would be considerably enhanced, perhaps even to a degree that it would then be possible to draw conclusions with absolute certainty about the past and the future fate of our own solar system, including the Earth.

Besides their immediate value, all of these research results would also have, however, the greatest significance for the future development of space travel, because when the conditions in those regions of space and on those celestial bodies at which our travel is aiming are exactly known to us, then a trip to outer space would no longer venture into the unknown, and therefore would lose some of its inherent danger.

A Giant Floating Mirror

The potentials of a space station are by no means exhausted with the above descriptions. Based on the condition that for the space station the sun shines unattenuatedly and continuously (disregarding occasional brief passes through the Earth's shadow), benefits could be derived, furthermore, for some technical applications on Earth. From the space station, the sun's radiation—even on a large scale—could be artificially focused on various regions of the Earth's surface if, as Oberth suggests, giant mirrors were erected that were appropriately built, orbited the Earth in a free orbital path, and hence were suspended above it.

According to Oberth, these mirrors would consist of individual segments, moveable in such a manner that any arbitrary orientation in the plane of the mirror can be remotely assigned to them through electrical signals. By appropriately adjusting the segments, it would then be possible, depending on the need, to spread the entire solar energy reflected by the mirror over wide regions of the Earth's surface or to concentrate it on single points, or finally to radiate it out into space if not being used.

"Space mirrors" of this type would be in a weightless state as a result of their orbital motion; this fact would considerably simplify their manufacture. According to Oberth, a circular network of wires could serve as a frame for their construction and, to this end, could be extended in space through rotation. The individual segments would be attached to the wire mesh and would consist of paper-thin sodium foils. According to Oberth's plans, a mirror of this type with a diameter of 100 km would cost around 3 billion marks and require approximately 15 years for its completion.

Besides this proposal, there would, no doubt, be still other possibilities of constructing a large floating mirror of this type. At smaller diameters of perhaps only several 100 meters, we could certainly succeed in giving the entire mirror such a rigid structure that it could be rotated at will around its center of mass, even in its entirety, by means of control motors, and that arbitrary positional changes could be performed with it.

The electrical energy necessary for controlling mirrors of this type would be available in the space station in sufficient quantity. The actual controls would have to be placed in the observatory and positioned in such a fashion that they could be operated at the same time while performing observations with the giant telescope, making it possible to adjust the mirrors' field of light precisely on the Earth.

The uses of this system would be numerous. Thus, important harbors or airports, large train stations, even entire cities, etc. could be illuminated during the night with natural sunlight, cloud cover permitting. Imagine the amount of coal saved if, for example, Berlin and other cosmopolitan centers were supplied with light in this fashion!

Using very large space mirrors, it would also be possible, according to Oberth, to make wide areas in the North inhabitable through artificial solar radiation, to keep the sea lanes to Northern Siberian harbors, to Spitzbergen, etc. free of ice, or to influence even the weather by preventing sudden drops in temperature and pressure, frosts, hail storms, and to provide many other benefits.

The Most Dreadful Weapon

But like any other technical achievement the space mirror could also be employed for military purposes and, furthermore, it would be a most horrible weapon, far surpassing all previous weapons. It is well known that fairly significant temperatures can be generated by concentrating the sun's rays using a concave mirror (in a manner similar to using a so-called "burning glass"). Even when a mirror has only the size of the human hand, it is possible to ignite a hand-held piece of paper or even wood shavings very simply in its focus (Figure 95).

Figure 95. Igniting a piece of wood using a concave mirror.

Key: 1. Sun's rays.

Imagine that the diameter of a mirror of this type is not just 10 cm, but rather several hundreds or even thousands of meters, as would be the case for a space mirror. Then, even steel would have to melt and refractory materials would hardly be able to withstand the heat over longer periods of time, if they were exposed to solar radiation of such an enormous concentration.

Now, if we visualize that the observer in the space station using his powerful telescope can see the entire combat area spread out before him like a giant plan showing even the smallest details, including the staging areas and the enemy's hinterland with all his access routes by land and sea, then we can envision what a tremendous weapon a space mirror of this type, controlled by the observer in orbit, would be!

It would be easy to detonate the enemy's munitions dumps, to ignite his war material storage area, to melt cannons, tank turrets, iron bridges, the tracks of important train stations, and similar metal objects. Moving trains, important war factories, entire industrial areas and large cities could be set ablaze. Marching troops or ones in camp would simply be charred when the beams of this concentrated solar light were passed over them. And nothing would be able to protect the enemy's ships from being destroyed or burned out, like bugs are exterminated in their hiding place with a torch, regardless of how powerful the ships may be, even if they sought refuge in the strongest sea fortifications.

They would really be death rays! And yet they are no different from this life-giving radiation that we welcome everyday from the sun; only a little "too much of a good thing." However, all of these horrible things may never happen, because a power would hardly dare to start a war with a country that controls weapons of this dreadful nature.

To Distant Celestial Bodies

In previous considerations, we did not leave the confines of the dominant force of the Earth's attraction—its "territory in outer space," so to speak. What about the real goal of space flight: completely separating from the Earth and reaching more distant celestial bodies?

Before we examine this subject, a brief picture of the heavenly bodies is provided, seen as a future destination from the standpoint of space travel. In the first place, we must broaden the scope of our usual notions, because if we want to consider the entire universe as our world, then the Earth, which previously appeared to us as the world, now becomes just our "immediate homeland." Not only the Earth! But everything that it holds captive by virtue of its gravitational force, like the future space station; even the Moon must still be considered a part of our immediate homeland in the universe, a part of the "Earth's empire." How insignificant is the distance of about 380,000 km to the Moon in comparison to the other distances in outer space! It is only a thousandth of the distance to Venus and Mars, located next to us after the Moon, and even the Earth together with the Moon's entire orbit could easily fit into the sun's sphere.

The next larger entity in the universe for us is the solar system, with all its various, associated heavenly bodies. These are the 8 large planets or "moving" stars, one of which is our Earth, (Figures 96 and 97) and numerous other celestial bodies of considerably smaller masses: plan-

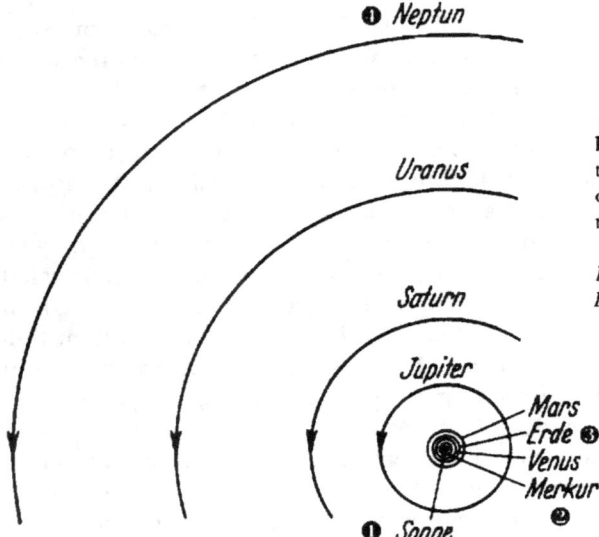

Figure 96. A schematic of the orbits of the 8 planets of our solar system in their relative sizes.

Key: 1. Sun; 2. Mercury; 3. Earth; 4. Neptune.

etoids, periodic comets, meteor showers, etc. Of the planets, Mercury is closest to the sun, followed by Venus, the Earth, Mars, Jupiter, Saturn, Uranus and Neptune, the most distant.* Together with the Moon, Venus and Mars are the planets nearest to the Earth.

All of these celestial bodies are continuously held captive to the sun by the effect of gravity; the bodies are continually forced to orbit the sun—as the central body—in elliptical orbits. The planets together with the sun form the "sun's empire of fixed stars," so to speak. They form an island in the emptiness and darkness of infinite space, illuminated and heated by the sun's brilliance and controlled at the same time by the unshakable power of the sun's gravitational force, and are thus

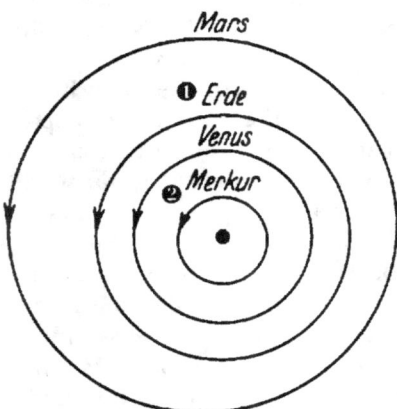

Figure 97. Enlarged rendition taken from Figure 96 of the orbits of Mars, the Earth, Venus and Mercury.

Key: 1. Earth; 2. Mercury.

* Note the absence of Pluto, which was not discovered until 1930.

linked in an eternal community. That island is our "extended homeland" within the universe. A realm of truly enormous size: even light needs more than 8 hours to traverse it and it is racing through space at a velocity of 300,000 km per second!

And yet, how tiny is this world compared to the incomprehensible distances of the universe, from which those many white hot celestial bodies familiar to us as fixed stars send their shining greetings of radiation. Even the one closest to us, the fixed star Alpha Centauri, is 4.3 light years away; i.e., around 4,500 times as far as the diameter of the entire solar system! All of the others are still many more light years away from us, most of them hundreds and thousands of light years. And if there are fixed stars closer to us that are already burnt out, then we are unaware of them in the eternal darkness of empty space.

From this discussion, it can be seen that only those heavenly bodies belonging to the solar system can be considered for the trip to alien celestial bodies, at least according to the views held today.

The Technology of Space Travel

Just exactly how the long trip through outer space can be achieved has already been indicated at the beginning of this book:[28] in general, in free orbits around those celestial bodies in whose gravitational field the trip is proceeding. Within its realm, the sun must consequently be continually orbited in some free orbit if a space ship is to avoid falling victim to its gravitational force and crashing into its fiery sea.

However, we do not have to take any special precautions as long as we stay close to the Earth or to another heavenly body of the solar system. After all, these bodies orbit the sun in their own free orbits, as do all bodies belonging to it. At the velocity of the Earth (30,000 meters per second), the Moon, for example, also circles the sun, as will our future space station (both as satellites of the Earth). As a result, the sun's gravitational force loses its direct effect on those two satellites ("stable state of floating" compared to the sun).

Only when the space ship moves further away from the immediate gravitational region of a celestial body circling the sun would the sun have to be orbited in an independent free orbit. If, for example, the trip is to go from the Earth to another planet, then, based on previous calculations, both the course of this independent orbit and the time of departure from the Earth must be selected in such a fashion that the space ship arrives in the orbit of the destination planet approximately at the time when the planet also passes through the encounter point.

28. See pages 8-9.

If the space vehicle is brought in this fashion into the practical effective range of gravity of the destination celestial body, then the possibility exists either to orbit the body in a free orbit as a satellite as often as desired or to land on it. Landing can, if the celestial body has an atmosphere similar to that of the Earth, occur in the same fashion as previously discussed[29] for the Earth (Hohmann's landing manoeuver, Figures 44 and 45). If, however, a similar atmosphere is absent, then the landing is possible only by reaction braking, that is, by operating the propulsion system opposite to the direction of free fall during landing[30] (Figure 37).

To travel to another celestial body within the solar system after escape from the original body, the orbital motion, previously shared with this body around the sun, must be altered by using the propulsion system to such an extent that the space ship enters an independent orbit around the sun, linking its previous orbit with that of the other celestial body. To implement this in accordance with the laws of celestial mechanics, the original orbital movement would have to be accelerated if the vehicle (according to the position of the target body) is to move away from the sun (Figure 98), and to be decelerated if it is to approach it. Finally, as soon as the destination is reached, the motion maintained in the "transfer orbit" must be changed into the motion that the vehicle must have as the new celestial body for effecting the orbiting or landing maneuver. The return trip would occur in the same fashion. It can be seen that repeated changes of the state of motion are necessary during a long-distance trip of this nature through planetary space. The changes would have to be produced through propulsion with an artificial force and, therefore, would require an expenditure of propellants, a point previously mentioned at the beginning of this book.[31] As determined mathematically by Hohmann, the propellant expenditure reaches a minimum when the orbits of the original celestial body and that of the destination body are not intersected by the transfer orbit of the vehicle, but are tangential to it (touch it) (Figure 99). Nevertheless, the required amounts of propellant are not insignificant.

Besides the points discussed above, there are additional considerations if the destination heavenly body is not to be orbited, but is supposed to be landed on. These considerations are all the more important the greater the mass and consequently the gravitational force of the destination planet are, because the re-ascent from the destination planet when starting the return trip requires, as we already know from the discussion of the Earth,[32] a very significant expenditure of energy. Additionally, if brak-

29. See pages 56-60.
30. See pages 53-54.
31. See page 8.
32. See pages 35-36.

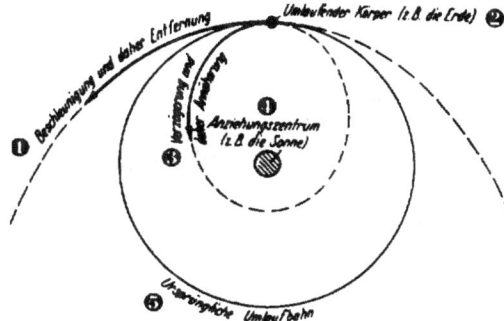

Figure 98. If the motion of a freely orbiting body is accelerated, then it expands its original orbit and moves away from the center of gravity. If the motion is decelerated, then the body approaches the center of gravity by contracting its orbit.

Key: 1. Acceleration and, therefore, increasing distance; 2. Orbiting body (e.g., the Earth); 3. Decelerating and, therefore, approaching; 4. Center of gravity (e.g., the sun); 5. Original orbit.

ing must be performed during the landing by propulsion in the absence of an appropriate atmosphere (reaction braking), then a further, significant increase of the amount of necessary propellants results.

The propellants must be carried on board from the Earth during the outward journey, at least for the initial visit to another planet, because in this case we could not expect to be able to obtain the necessary propellants from the planet for the return trip.

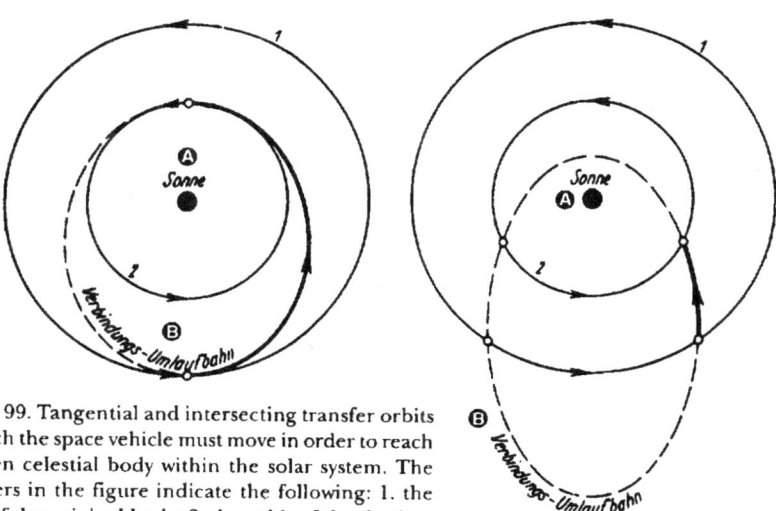

Figure 99. Tangential and intersecting transfer orbits in which the space vehicle must move in order to reach an alien celestial body within the solar system. The numbers in the figure indicate the following: 1. the orbit of the original body; 2. the orbit of the destination celestial body. The distance of the transfer orbit marked by heavy lines is that part of the orbit which the vehicle actually travels through.

Key: A. Sun; B. Transfer orbit.

Launching from the Earth's Surface

If a trip of this nature were launched directly from the Earth's surface, this entire amount of propellant would have to be first separated from the Earth (by overcoming its gravitational force). According to what was stated previously,[33] an extraordinary expenditure of energy is necessary for this purpose.

For the present case, at least with the efficiencies of currently available propellants, the amount to be carried on board would constitute such a high fraction of the total weight of the vehicle that it could hardly be built.

The only visit to a celestial body that could probably be undertaken directly from the Earth's surface with propellants known to date, would be an orbiting of the Moon for exploring its surface characteristics in more detail, in particular, the side of the Moon that continually faces away from the Earth. During this trip, the space ship could also be "captured" by the Moon, so that it would circle the Moon as often as necessary in a free orbit as a moon of the Moon. The amount of propellants necessary for this effort would not be much greater than for a normal ascent from the Earth up to escape velocity.

The Space Station as a Base for Travel into Deep Space

The conditions, however, would be considerably more favorable if a depot for propellants appropriately suspended high over the Earth and continuously circling it in a free orbit was built, as Oberth suggests, and if the trip was started from this depot instead of from the Earth's surface, because in that case only a modest expenditure of energy would be necessary for a complete separation from the Earth, and the vehicle need not, therefore, be loaded with the propellants necessary for the ascent from the Earth. It would have to carry on board only slightly more than the amount necessary for the deep space trip itself.

Since the depot would be in a weightless state as a result of its free orbital motion, the propellants could simply be stored there freely suspended in any amount and at any place. Protected against the sun's rays, even oxygen and hydrogen would remain solidly frozen for an indefinite time. Their resupply would have to be accomplished by a continuous space ship shuttle service either from the Earth where the propellants (at least liquid oxygen and hydrogen) could be produced, for example, in large power plants powered by the heat of the tropical seas; or from

---

33. See pages 35-36.

the Moon, as Max Valier suggests. This method would be particularly advantageous, because since the mass and consequently the gravitational force of the Moon are considerably smaller than those of the Earth, the expenditure of energy necessary for the ascent and consequently for the propellant supply for that ascent would be significantly less. However, this assumes that the required raw materials would, in fact, be found on the Moon, at least water (in a ice-like condition, for instance) because it can be decomposed electrolytically into oxygen and hydrogen, the energy for this process being provided by a solar power plant. Unfortunately, the probability for this is not all that high.

If, however, this should be possible, then even the Moon, according to Hohmann's recommendation, could be used as a starting point for travel into deep space; that is, the propellant depot could be built on the Moon. Despite many advantages of this idea, Oberth's recommendation of a freely suspended depot appears to be the better one, because the complete separation from the gravitational field of the Earth (including the Moon) would require considerably less expenditure of energy from a depot of this nature. More specifically, it would certainly be the most advantageous from an energy economics point of view to build the depot one or more millions of kilometers away from the Earth, especially if the propellants must be supplied from the Earth. We want, however, to build the depot at our space station, and thus make it a transportation base, because it is already equipped with all facilities necessary for this purpose.

Of this equipment, giant telescopes, among others, would be particularly valuable because thanks to their almost unlimited capabilities they would not only make it possible to study in detail the celestial destinations from a distance, a point previously described.[34] The space station could probably keep the space ship under constant surveillance during a large part of its trip, in many cases perhaps even during the entire trip, and could remain in at least one-way communications with it through light signals to be emitted at specific times by the space ship. Thus, the space station, besides satisfying the many assignments already discussed, would be able to satisfy those that assist not only in preparing for actual travel into the universe but also serve as a basis for the entire traffic into outer space.

The Attainability of the Neighboring Planets

Hohmann has studied in detail the problem of travelling to other celestial bodies. According to his results, the long-distance trip would

---

34. See pages 120-121.

last 146 days from the Earth to Venus and 235 days to Mars, expressed in a terrestrial time scale. A round trip including a flyby of both Venus and Mars at the relatively small distance of approximately 8 million kilometers could be carried out in about 1.5 years. Almost 2.25 years would be necessary for a visit to Venus with a landing, including a stay there of 14.5 months and the two-way travel time.

Assume now the following: in the sense of our previous considerations, the trip would start from the space station, so that only a modest amount of energy would be necessary for the complete separation from the Earth's gravitational field; the return trip would take place directly to the Earth's surface, so that no propulsive energy would have to be expended, because in this case the descent could be controlled by using only air drag braking. The load to be transported would be as follows: 2 people including the supplies necessary for the entire trip, and all instruments required for observation and other purposes.

It then follows from Hohmann's calculations that the vehicle in a launch-ready condition, loaded with all propellants necessary for traveling there and back, would have to weigh approximately the following: 144 tons for the described round trip with a flyby near Venus and Mars, of which 88% would be allocated to the propellants, 12 tons for the first landing on the Moon, 1350 tons for a landing on Venus and 624 tons for a Mars landing. For the trip to the Moon, 79% of the entire weight of the vehicle would consist of the propellants carried on board, but approximately 99% for the trips to Venus and Mars. A 4,000 meter per second exhaust velocity was assumed in these cases.

It is obvious that the construction of a vehicle that has to carry amounts of propellants on board constituting 99% of its weight would present such significant engineering difficulties that its manufacture would initially be difficult to accomplish. For the present, among our larger celestial neighbors, only the Moon would, therefore, offer the possibility of a visit with a landing, while the planets could just be closely approached and orbited, without descending to them. Nevertheless one can hope that we will finally succeed in the long run—probably by employing the staging principle explained in the beginning[35]—even with technologies known today in building space rockets that permit landings on our neighboring planets.

With the above, and when considering the present state of knowledge, all possibilities are probably exhausted that appear to present themselves optimistically for space ship travel. The difficulties would be much greater confronting a visit to the most distant planets of the solar system. Not only are the distances to be travelled to those destinations much

---

35. See pages 37-39.

longer than the ones previously considered, but since all of these celestial bodies have a far greater distance from the sun than the Earth, the sun's gravitational field also plays a significant role in their attainability. Because if, for example, we distance ourselves from the sun (i.e, "ascend" from it), then in the same fashion as would be necessary in the case of the Earth's gravitational field, the sun's gravitational field must be overcome by expending energy, expressed as the change of the orbital velocity around the sun, and the distance from its center, as previously discussed.[36] This is required in long-distance travels throughout planetary space.

If, however, we also wanted to descend down to one of these celestial bodies, then enormously large amounts of propellants would be necessary, in particular for Jupiter and Saturn because they have very strong gravitational fields as a result of their immense masses. In accordance with the above discussion, we naturally cannot even think of reaching the fixed stars at the present time, solely because of their enormous distance.

Distant Worlds

This doesn't mean to say that we must remain forever restricted to the Earthly realm and to its nearest celestial bodies. Because, if we succeeded in increasing further the exhaust velocity beyond the 4,000 (perhaps 4,500) meters per second when generating the thrust, the highest attainable in practice at the present time, or in finding a possibility of storing on board very large quantities of energy in a small volume, then the situation would be completely different.

And why shouldn't the chemists of the future discover a propellant that surpasses in effectiveness the previously known propellants by a substantial degree? It might even be conceivable that in the course of time mankind will succeed in using those enormous amounts of energy bound up in matter, with whose presence we are familiar today, and in using them for the propulsion of space vehicles. Perhaps we will someday discover a method to exploit the electrical phenomenon of cathode radiation, or in some other way attain a substantial increase of the exhaust velocity through electrical influences. Even using solar radiation or the decay of radium, among others, might offer possibilities to satisfy this purpose.

In any case, natural possibilities for researchers and inventors of the future are still available in many ways in this regard. If success results from these efforts, then probably more of those alien worlds seen by us

36. See pages 128-129.

only as immensely far away in the star-studded sky could be visited by us and walked on by humans.

An ancient dream of mankind! Would its fulfillment be of any use to us? Certainly, extraordinary benefits would accrue to science. Regarding the practical value, an unambiguous judgement is not yet possible today. How little we know even about our closest neighbors in the sky! The Moon, a part of the Earthly realm, our "immediate homeland" in the universe, is the most familiar to us of all the other celestial bodies. It has grown cold, has no atmosphere, is without any higher life form: a giant rock-strewn body suspended in space, full of fissures, inhospitable, dead—a bygone world. However, we possess significantly less knowledge about that celestial body, observed the best next to the Moon, about our neighboring planet Mars, even though we know relatively much about it in comparison to the other planets.

It is also an ancient body, although considerably less so than the Moon. Its mass and, consequently, its gravitational force are both considerably smaller than that of the Earth. It has an atmosphere, but of substantially lower density than the terrestrial one (the atmospheric pressure on its surface is certainly significantly lower than even on the highest mountain top on Earth). Even water is probably found on Mars. However, a fairly large part of it is probably frozen, because the average temperature on Mars appears to be substantially below that of the Earth, even though in certain areas, such as in the Martian equatorial region, significantly warmer points were detected. The temperature differences between night and day are considerable due to the thinness of the atmosphere.

The most unique and most frequently discussed of all Martian features are the so-called "Martian canals." Even though in recent times they have been considered mostly as only optical illusions, it is still unclear just exactly what they are.

In any case, the present knowledge about Mars does not provide sufficient evidence for a final judgement as to whether this celestial body is populated by any form of life, or even by intelligent beings. For people from Earth, Mars would hardly be inhabitable, primarily because of the thinness of its atmosphere. From a scientific point of view, it would certainly offer an immensely interesting research objective for space travelers. Whether walking on Mars would have any practical value can still not be determined with certainty today; however, this does not appear to be very probable.

It is an altogether different situation with the second planet directly adjacent to us, Venus, the brightly shining, familiar "morning and evening star." Its size as well as its mass and accordingly the gravitational field existing at its surface are only slightly smaller than the Earth's. It also has an atmosphere that should be quite similar to the terrestrial atmosphere, even though it is somewhat higher and denser than the Earth's. Unfortunately, Venus can be observed only with difficulty from the Earth's surface, because it is always closer to the sun and, therefore, becomes visible only at dawn or dusk. As a result, we know very little about its rotation. If Venus rotates in approximately 24 hours roughly like the Earth, a situation assumed by some experts, then a great similarity should exist between Venus and Earth.

In the case of this planet, finding conditions of life similar to terrestrial conditions can be expected with high probability, even if the assumption should be valid that it is continually surrounded by a cloud cover. Because even on Earth, highly developed forms of plant and animal life already existed at a time when apparently a portion of the water now filling the seas and oceans was still gaseous due to the slow cooling of the globe millions of years ago and, therefore, continually surrounded our native planet with a dense cloud cover. In any case, Venus has the highest probability of all the celestial bodies closely known to us of being suitable for colonization and, therefore, of being a possible migration land of the future. Furthermore, since it is nearest to us of all planets, it could be the most likely and tempting destination for space travel.

Mercury offers even more unfavorable conditions for observation than Venus because it is still closer to the sun. Of all the planets it is the smallest, has an atmosphere that is no doubt extremely thin and surface conditions apparently similar to those of the Moon. For this reason and especially due to its short perihelion distance (solar radiation about 9 times stronger than on the Earth!), extremely unfavorable temperature conditions must exist on it. Consequently, Mercury should be considerably less inviting as a destination.

While it was still possible when evaluating the celestial bodies discussed above to arrive at a fairly probable result, our current knowledge about the more distant planets, Jupiter, Saturn, Uranus and Neptune, is hardly sufficient to achieve this. Although we have been able to determine that all of them have dense atmospheres, the question of the surface conditions of these planets is, however, still entirely open: in the cases of Jupiter and Saturn, because they are surrounded by products of condensation (clouds of some kind) so dense that we apparently cannot even see their actual surfaces; and in the cases of Uranus and Neptune, because their great distances preclude precise observation.

Therefore, anything regarding their value as a space flight destination can only be stated with difficulty. But the following condition by itself is enough to dampen considerably our expectations in this regard: a relatively very low average density has been determined for these planets ($1/4$ to $1/5$ of that of the Earth), a condition indicating physical characteristics quite different from those on Earth. It would be perhaps more likely that several of the moons of these celestial bodies (primarily, those of Jupiter would be considered in this connection) offer relatively more favorable conditions. One thing is certain in any case: that their masses are considerably greater than the Earth's and that, therefore, the powerful gravitational fields of these planets would make a visit to them extraordinarily difficult, especially in the cases of Jupiter and Saturn.

Regarding the remaining, varying types of celestial bodies that still belong to the solar system, it can be said with a fair degree of certainty today that we would hardly be able to benefit in a practical sense from a trip to them. We see then that, generally speaking, we should not indulge in too great hopes regarding the advantages that could be derived from other celestial bodies of our solar system. In any case, we know far too little about them not to give free reign to the flight of thoughts in this regard:

Of course, it could turn out that all of these worlds are completely worthless for us! Perhaps, however, we would find on some of them a fertile soil, plant and animal life, possibly of a totally alien and unique nature for us, or perhaps of a gigantic size, as existed on Earth long ago. It would not be inconceivable that we would meet even humans or similar types of life, perhaps even with civilizations very different from or even older than those of our native planet. It is highly probable that life on other planets—if it exists there at all—is at another evolutionary stage than that on Earth. We would be able then to experience that wonderful feeling of beholding images from the development of our own terrestrial existence: current, actual, living and yet—images from an inconceivable, million-year old past or from an equally distant future.

Perhaps we would discover especially valuable, very rare Earthly materials, radium for example, in large, easily minable deposits? And if the living conditions found there are also compatible with long-term human habitation, then perhaps even other celestial bodies will one day be possible as migration lands—regardless of how unbelievable this may sound today. That such planets exist among those of our solar system is, however, only slightly probable according to what has been stated previously, with the exception of Venus, as already noted.

Will It Ever be Possible to Reach Fixed Stars?

It would be much more promising, however, if the stars outside of our solar system could also be considered in this context, because the number is enormous even of only those celestial bodies that, since they are in a white-hot state, are visible and, therefore, are known to us as fixed stars. Many of these are similar to our sun and, as powerful centers of gravity, are probably orbited exactly like the sun by a number of small and large bodies of varying types.

Shouldn't we expect to find among these bodies also some that are similar to our planets? Of course, they are too far away for us to perceive them; however, probability speaks strongly for their existence. In fact, the most recent scientific research—as one of its most wonderful results—has been able to show that the entire universe, even in its most distant parts, is both controlled by the same natural laws and structured from the same material as the Earth and our solar system! At other locations within the universe, wouldn't something similar, in many cases almost the same thing, have to materialize under the same conditions (from the same matter and under the effect of the same laws) as in our case?

It is certainly not unjustified to assume that there would be other solar systems more or less similar to ours in the universe. And among their numerous planets, there surely would be some that are almost similar to the Earth in their physical and other conditions and, therefore, could be inhabited or populated by people from Earth, or perhaps they may already be populated by some living beings, even intelligent ones. At least the probability that this may be the case is significantly greater than if we only consider the relatively few planetary bodies of our solar system.

Yet, would it really be conceivable that those immeasurable distances still separating us even from the closest fixed stars could be traveled by humans, even taking into account the limit that is set by the average life span of a person, completely ignoring the related necessary technical performance of the vehicle?

Let's assume that a goal, which appears enormous even for today's concepts, has been achieved: perfecting the rocket propulsion system to such an extent that an acceleration of approximately 15 m/sec^2 could continually be imparted to the space ship over a very long time, even through years. Humans would probably be able to tolerate this acceleration over long periods of time through a gradual accommodation. To travel a given distance in space, it would then be possible to accelerate the vehicle continually and uniformly over the entire first half of its trip, that is, to give it more and more velocity, and to decelerate it in the same way over the second half and consequently to brake it gradually again (Figure 100). With this method, a given distance will be covered in the shortest possibly achievable time with given constant acceleration and deceleration.

If the trip now took place to neighboring fixed stars in this manner, then the following travel time would result for the entire round trip (the first visit would have to be a round trip) based on mathematical calculations: 7 years to Alpha Centauri, the star known to be the closest to us, and 10 years to the four fixed stars next in distance; numerous fixed stars could be reached in a total round trip travel time of 12 years.

However, it is quietly assumed here that any velocity, without limitation, is possible in empty ether space. In accordance with the theory of relativity, a velocity greater than the speed of light of 300,000 km per second can never be attained in nature.

If this is taken into account and if it is further assumed that no other obstacle (currently unknown to us, perhaps inherent in the nature of universal world ether) would prevent us from attaining travel velocities approaching the speed of light, then we could, nevertheless, reach the fixed star Alpha Centauri in around 10 years, the four farther stars in 20 years, and a considerable number of neighboring fixed stars presently known to us in 30 years; these figures represent total round trip travel times.

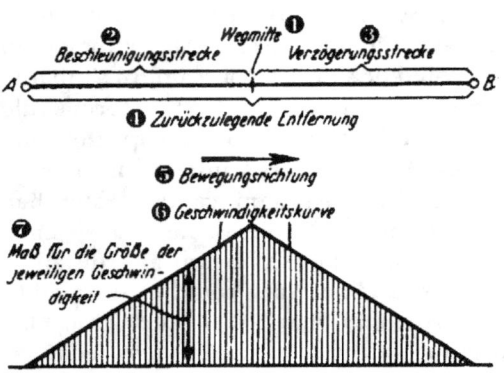

Figure 100. Covering a distance when the vehicle is uniformly accelerated over the entire first half of the distance, and similarly the vehicle is decelerated over the second half. The highest velocity of motion resulting from this method is reached at the mid-point.

Key: 1. Mid-point; 2. Accelerated over this distance; 3. Decelerated over this distance; 4. Distance to be traveled; 5. Direction of motion; 6. Velocity curve; 7. Instantaneous velocity.

For the one-way trip, which would be of interest for continual traffic, half as much time would suffice. No doubt, trips of such duration would be fairly close to the limit of human endurance; however, they cannot yet be discarded as completely non-implementable, since no fundamental obstacle can, in fact, be seen for reaching the closest fixed stars.

Meanwhile, the question still remains open as to whether vehicles could ever be built having the technical perfection necessary for such performances. However, even this question cannot be answered with an unequivocal no because, as has been pointed out previously,[37] natural phenomena exist that could provide possibilities, such as exploiting the energy bound up in matter by smashing atoms, or utilizing the decay of radium, or cathode radiation, etc.

Admittedly, we are far away today from that goal of completely mastering such natural phenomena to such an extent that we would be able

---

37. See page 131.

to use them in an engineering sense for space travel purposes! And, we don't know whether this will ever be successful at all.

As far as is humanly possible to predict, the sons of our time will hardly achieve this. Therefore, the fixed stars, which conceal the great secrets of the universe in their immensity, will no doubt remain unreachable for them. Who can say what scientific triumphs and technical potentials future times will bring! Since mankind has now become confident with scientific reasoning, what tremendous progress is achieved today in only a few decades; and what are a hundred, even a thousand years in those eons of human development still ahead of us.

Conquering space! It would be the most grandiose of all achievements ever dreamed of, a fulfillment of the highest purpose: to save the intellectual accomplishments of mankind for eternity before the final plunge into oblivion. Only when we succeed in transplanting our civilization to other celestial bodies, thus spreading it over the entire universe, only when mankind with all its efforts and work and hopes and with what it has achieved in many thousands of years of striving, only when all of this is no longer just a whim of cosmic events, a result of random incidents in eternal nature's game that arise and die down with our little Earth—so large for us and yet so tiny in the universe—will we be justified to feel as if we were sent by God as an agent for a higher purpose, although the means to fulfill this purpose were created by man himself through his own actions.

The Expected Course of Development of Space Travel

Now let us turn back from these dreams of the future to the reality of the present. It would really be an accomplishment today if we succeeded in lifting an unmanned rocket several 10s or even 100s of kilometers! Even though the problems associated with space travel have been worked out theoretically to some degree thanks to the manyfold efforts of the last few years, almost everything still has to be accomplished from a practical standpoint. Therefore, at the conclusion of this book, possible directions of space travel development are briefly outlined.

The first and most important point in this regard is, without a doubt, the engineering improvement of the rocket engine, the propulsion system of the space ship. This is a task that can be solved only in thorough, unselfish research. It is a problem that should be worked out first and foremost in the experimental laboratories of universities and on the test fields of experienced machine factories.

In connection with the above, experience must be accumulated (at least as far as space rockets for liquid propellants are concerned) in the methods of handling liquefied gases, in particular liquid oxygen, and liquid hydrogen, among others. Furthermore, the behavior of metals at extremely low temperatures should be tested in laboratory experiments in order to determine the substances best suited as construction materials for space ships. Finally, the method of manufacturing propellant tanks will also require detailed studies.

After solving these fundamental engineering issues, the following could then be considered next: to launch unmanned space rockets into the higher layers of the atmosphere or even above them into empty outer space and to let them descend using a parachute, as far as the latter turns out to be achievable.

These experiments will make it possible not only to accumulate the necessary technical experiences concerning rocket technology, but in particular to become familiar also with the laws of aerodynamics at abnormally high velocities and of the laws of heating due to atmospheric friction, data that are of utmost importance for shaping the vehicle itself as well as the parachutes, wings, etc. We will furthermore be able to determine up to what altitudes simple parachute landings are still feasible (taking into consideration the danger of burning the parachute due to atmospheric friction). As a result of these experiments, exact information can finally be obtained about the nature of the higher layers of the Earth's atmosphere, knowledge that forms a most important basis for the further development of space travel, but would also be of great value in many other regards (radio technology, for example).

Firing an unmanned space rocket loaded with flash powder at the Moon, as recommended by many parties, could probably also be attempted as a subsequent step; this would have very little practical value, however.

In parallel with these efforts, we—in order to prepare for the ascent of humans—would have to research the physical tolerance of elevated gravitational effects by performing appropriate experiments using large centrifuges (or carousels) and, furthermore, to create the possibility for remaining in airless space by perfecting the previous methods of supplying air artificially and by testing appropriate space suits in containers made airless and cooled to very low temperatures.

As soon as the results of the previously delineated, preparatory work allow, ascents using simple parachute landings can then be carried out (possibly following previous launchings with test animals) by means of manned space rockets up to altitudes determined beforehand as reliable for such flights. Now we can proceed to equip the vehicles with wings to make them capable of gliding flight landings (Hohmann's landing manoeuver) and consequently for attaining those altitudes from which a simple parachute landing would no longer be feasible.

Experience in the engineering of the rocket propulsion system necessary for building such airplane-like space ships (or expressed in another way: airplanes powered by rocket, that is "recoil airplanes," "rocket airplanes," etc.) and experience with atmospheric friction, air drag, etc. will both have been gained at this time from the previously described preliminary experiments made with unmanned space rockets.

When testing these vehicles, which would be performed by using as extensively as possible previous experiences with aviation, we will first start with relatively short flight distances and altitudes and try to increase these distances and altitudes, gradually at first, then more and more through a corresponding increase of the flight velocities.

As soon as maneuvering with rocket airplanes in general and especially the flight technology necessary at cosmic velocities in the higher, thin layers of air are mastered, the following achievements are practically automatic:

1. Creating terrestrial "express flight transportation at cosmic velocities," as explained in the beginning of the book; that is, the first practical success of space flight is attained (every ascent of this type not flown above the atmosphere with a gliding flight landing is strictly speaking nothing other than an express flight of this nature);

2. Making possible the fact that returning space ships can now descend using a gliding flight landing (instead of a simple parachute landing); i.e., the safe return to Earth from any arbitrary altitude is assured as a result, an accomplishment that is of the greatest importance for space flight and signifies an essential precondition for its implementation.

This previously described course of development (first, performing ascents using unmanned space rockets with a simple parachute landing and, only on the basis of the experiences gained during these ascents, developing the rocket airplane) would presumably be more practical than developing this airplane directly from today's airplane, as has been advocated by others, because experiences to be accumulated initially during this development will probably force a certain method of construction for the rocket airplane that may differ considerably from the methods used for airplanes employed today. To arrive, however, at this probable result solely through experiments with airplanes (which are costly), would presumably be significantly more expensive and moreover entail much more danger.

In any case, the most important point is that practical experiments are started as soon as possible. By a gradual increase in the performance of rocket airplanes or airplane-like space ships, more significant horizontal velocities and altitudes will finally be attained in the course of time, until finally free orbital motion above the atmosphere and around the Earth will result. Arbitrarily selecting the orbit will no longer present any difficulties.

Then the potential for creating the previously described space station, that is, achieving the second practical success in the development of space travel, is already given. Also, random high ascents could now be undertaken, and the Moon could eventually be orbited.

Both express flight transportation and the space station are purely terrestrial matters. Now we will strive to realize the additional goals of space flight while using the space station as a transportation control point: walking on the Moon, if possible building a plant on the Moon for producing propellants, orbiting neighboring planets, and other activities that may prove feasible.

Final Remarks

Even if, contrary to expectations, we were not successful soon in attaining in a practical manner the higher exhaust velocities necessary for the goals mentioned above by using sufficiently simple systems, and even if the exhaust velocity could be raised only up to about 2,000-3,000 meters per second, then space flight would nevertheless give us in the near future the ability to research thoroughly the Earth's atmosphere up into its highest layers, and especially—as a direct practical benefit—to create the described terrestrial express flight transportation at cosmic velocities, until later times will finally bring the realization of the other goals.

By just accomplishing the above goals, a success would be achieved that would far overshadow everything previously created in the technical disciplines. And it really can no longer be doubted that this would at least be achievable even today with a determined improvement of available engineering capabilities. This will be successful that much sooner the earlier and with the more thorough and serious scientific effort we tackle the practical treatment of the problem, although we must not underestimate the extent of the difficulties that still have to be overcome.

However, the purpose of the present considerations is not an attempt to convince anyone that we will be able tomorrow to travel to other celestial bodies. It is only an attempt to show that traveling into outer space should no longer be viewed as something impossible for humans but presents a problem that really can be solved by technical work. The overwhelming greatness of the goal should make all the roadblocks still standing in its way appear insignificant.

Index

absolute zero, 93
aerospace plane, see rocket as airplane
air, absence of, 88-89
air lock, 94-95
air supply for space station, 97
air resistance (drag), 9-10, 12, 32, 42
  braking through use of, 54-55, 58-59, 63-64, 71, 116
  heat resulting from, 65, 116
  minimizing, 70
Alpha Centauri, 125, 135-136
ascent, 25-32, 125-128
astronomy, 90, 120-121
atmospheric interference with observations, 90
atomic energy, 131, 136

ballistic trajectory, 8, 50
balloon for launching, 49, 50
Bergerac, Cyrano de, 16
booster rocket, 43, 50
braking ellipses, 59-61

cable for habitat wheel, 106-107
cartography, 119
center of gravity, 13, 101
centrifugal force, 4, 7, 57-58, 60, 66, 70, 72-78, 84, 101
centrifuge, 76, 138
climbing velocities, 43-44
combustion chamber, 33-36, 45, 47
communication, 98-99, 112
control and stabilization, 43, 49, 99-101
cooling of combustion chamber, etc., 34-35, 45, 51
costs, 47-48
cross-sectional loading, 42
cryogenics, 52

descent to Earth, 53-61, 125-128
dirigibles for launching, 43

E region of the ionosphere, 99n
Earth's atmosphere, 9-10
efficiency of the rocket vehicle, 19-29, 46-47
efficiency of the rocket motor, 40-41, 47
Einstein, Albert, 49
electrical propulsion, 131

energy content, expenditure of, 30-31, 51
escape velocity, 9, 11, 32, 38, 47, 52, 53
exhaust velocity, see velocity of expulsion
extraterrestrial life, 134-135

fireworks rocket, 15, 33
food service and eating in a weightless state, 87-88
free orbit, 6-8, 31, 57-61, 72-73, 125, 128
Gail, Otto Willi, 17
Ganswindt, Hermann, 16
Goddard, Robert H., 17, 37, 40-41, 47, 50, 64
Golightly, Charles, 16
gravity, 3-6, 12, 57-58, 75-78, 81, 101
  absence of, 80-88
gyroscope, 49

habitat wheel, 101-108
hammocks, 40, 115
heating of space station, 97-98
Heaviside layer, 99
Heaviside, Oliver, 99n
Hoeft, Franz Edler von, 17, 49-50
Hohmann, Walter, 17, 30, 48-49, 55-57, 60, 64, 65, 68, 69, 126, 129, 138

inclined trajectory, 62-65
inertial forces, 3-4, 75-78, 81, 85, see also centrifugal force
internal efficiency of rocket motor, 40-41, 47

Jupiter, 131, 133-134

Kennelly, Arthur E., 99n
Kennelly-Heaviside layer, 99
Keplerian ellipses, see ballistic trajectory
kinetic force, 18

laboratory for experiments, 110
landing, see descent to Earth, Hohmann
Lasswitz, Kurt (or Kurd), 16
light for space station, 96-97
lighting conditions in space, 89-90
Lorenz, Hans, 11

machine room, 101, 109, 111-112
mapping, see cartography

Mars, 130, 132
Mercury, 133
meteorology, 119-110
metallurgy, 39, 138
mirror for collecting Sun's rays, 112, 121-123
Moon, 123, 128-129, 130, 131

Neptune, 133
Nernst, Walther Hermann, 49
Newton, Isaac, 16
nozzle, 33-35, 45, 47

Oberth, Hermann, 11, 17, 30, 32, 37, 41-48, 52, 61, 75, 78-79, 97, 118, 121-122, 128, 129
observatory, 101, 109-111, 119-121
orbital inclination, 74
orbital velocity, 57, 73-74
overpressure in propellant tanks, 45

parachute for braking, 60-61
parachute for recovery, 43
passenger and storage compartment, 39-40, 43, 48
photography, 119
Pluto, 124n
propellants, 33-37, 40-46, 51-53, 68, 128, 131, 138
propellant injection, 34-35, 45, 51
propellant pumps, 45
propellant tanks, 35, 39, 45

reaction braking, 53-54, 61
reactive force, 12-15
recycling, 95
refrigeration, 96, 118
rocket, 15-17, 32-55, 115
 as airplane, 65-71, 139
 as weapon, 65

safety, 71
Saturn, 131, 133-134
silence in space, 89
solar power, 91-93, 95-96, 121-123
solar system, 123-125
space flight not utopian, 61
space mirror, see mirror
space station, 74-75, 78, 93-123, 128-129
space suits, 40, 82, 89, 95, 113-115, 138
space travel, 8-9, 115-118
spying by satellite, 120
stable state of suspension, 7-8, 78
staging principle, 37-39, 50
stratosphere, 71
structures for spacecraft, 88
submarine technology as applicable to space, 89

telescopes, see observatory
theory of relativity, 49n, 85, 136
thermal conditions in space, 91-93
travel velocity, 18-25, 31
troposphere, 71
Tsiolkovsky, Konstantin E., 16-17n

Uranus, 133

Valier, Max, 17, 129
velocity of expulsion, 20-23, 27-29, 32-38, 40, 41, 42, 46, 130, 131
ventilation for space station, 97-98
Venus, 130, 132-133
Verne, Jules, 16

water supply for space station, 98
water vessels in weightlessness, 86-87
weapon, 122-123
weather forecasting, see meteorology
weightlessness, 78-88
wings on a spacecraft, 57-59

Author of the Foreword

Frederick I Ordway III graduated from Harvard University in 1949 with a B.Sc. degreee in the geosciences, petroleum and mining. He did graduate work at a number of institutions including the Université de Paris and completed the Air University's curriculum in guided missiles. His professional career included employment with a number of industrial firms, the Army Ballistic Missile Agency, NASA's Marshall Space Flight Center, and the Department of Energy, where he directed the Special Projects Office until his retirement in 1994. He was a professor at the University of Alabama Research Institute and School of Graduate Studies and Research, and holds an honorary doctorate from the university. In addition, he is an aerospace writer of note. He has written or edited numerous books including, *History of Rocketry and Space Travel* (Thomas Y. Crowell and Harper & Row Co., four editions through 1985), with Wernher von Braun; *The Rocket Team* (Thomas Y. Crowell, 1979), with Mitchell Sharpe; and *Blueprint for Space: Science Fiction to Science Fact* (Smithsonian Institution Press, 1992), with Randy Liebermann. He has just completed the two-volume *Wernher von Braun: Crusader for Space* (Krieger, 1994), with Ernst Stuhlinger. His more than 250 articles have appeared in a number of places including the volumes in the *AAS History Series*, one of which he edited, and the *Journal of the British Interplanetary Society*, of which he occasionally serves as guest editor.

The Editors

Ernst Stuhlinger earned a Ph.D. in physics at the University of Tübingen with a dissertation on cosmic rays in 1936 and taught at the Technical University of Berlin from 1936 to 1941, also working with the German atomic energy program under Werner Heisenberg after 1939. He was drafted into the army in 1941 and in 1943 was transferred to the rocket development center at Peenemünde. He continued working under Wernher von Braun in the United States after World War II, first at Fort Bliss, Texas, and White Sands, New Mexico, and after 1950 for the Army's Ballistic Missile Agency and then NASA's Marshall Space Flight Center in Huntsville, Alabama. His work included guidance and control systems, satellite projects, scientific space studies, electric space propulsion, and other research programs. After serving as the associate director for science from 1968 to 1976, he retired from MSFC and taught astrophysics and space sciences at the University of Alabama in Huntsville. He has been the recipient of numerous awards and prizes including the Roentgen Prize, the Galabert Prize, the Hermann Oberth Award and Medal, the AIAA Propulsion Award, the Wernher von Braun Prize, and the Alexander von Humboldt Prize. The author of well in excess of 200 technical papers and articles published in journals such as *Weltraumfahrt* and the *Bulletin of the Atomic Scientists*, he has also written many books, including *Astronautical Engineering and Science* (McGraw-Hill, 1963); *Ion Propulsion for Space Flight* (McGraw-Hill, 1964); *Space Science and Engineering* (McGraw-Hill, 1965); and the two-volume *Wernher von Braun: Crusader for Space* (Krieger, 1994), with Fred Ordway.

J.D. Hunley earned a Ph.D. in history from the University of Virginia in 1973. After teaching for five years (1972-1977) at Allegheny College, he worked for more than a decade as a civilian historian in the Air Force History program and then became a historian in the NASA History Office in 1991. Among his awards have been election to Phi Beta Kappa at the University of Virginia; the C.S. Ashby Henry Prize in history; a Woodrow Wilson fellowship; and the Air Force History Program's monograph award. He has written or edited *Boom and Bust: Society and Electoral Politics in the Düsseldorf Area, 1867-1878* (Garland, 1987); *The Life and Thought of Friedrich Engels: A Reinterpretation* (Yale University Press, 1991); and *The Birth of NASA: The Diary of T. Keith Glennan* (NASA History Series, 1993). His articles have appeared in *Social Theory and Practice, Societas—A Review of Social History, Essays in History*, and *Comparative Bibliographic Essays in Military History*.

Jennifer Garland received her bachelor's degree in International Studies from the Honors College at Michigan State University in 1990 and has spent more than six years working in international information management and foreign languages, with particular focus on Eastern Europe, Russia, and the newly independent states in that region. At the time she helped edit this book, she was serving as Coordinator of Foreign Literature Services in the NASA Scientific and Technical Information Program. There, her primary responsibility was management of the NASA program to provide translation support into and out of more than 34 languages to U.S. Government agencies and contractors.

The NASA History Series

Reference Works, NASA SP-4000:

Grimwood, James M. *Project Mercury: A Chronology.* (NASA SP-4001, 1963).

Grimwood, James M., and Hacker, Barton C., with Vorzimmer, Peter J. *Project Gemini Technology and Operations: A Chronology.* (NASA SP-4002, 1969).

Link, Mae Mills. *Space Medicine in Project Mercury.* (NASA SP-4003, 1965).

Astronautics and Aeronautics, 1963: Chronology of Science, Technology, and Policy. (NASA SP-4004, 1964).

Astronautics and Aeronautics, 1964: Chronology of Science, Technology, and Policy. (NASA SP-4005, 1965).

Astronautics and Aeronautics, 1965: Chronology of Science, Technology, and Policy. (NASA SP-4006, 1966).

Astronautics and Aeronautics, 1966: Chronology of Science, Technology, and Policy. (NASA SP-4007, 1967).

Astronautics and Aeronautics, 1967: Chronology of Science, Technology, and Policy. (NASA SP-4008, 1968).

Ertel, Ivan D., and Morse, Mary Louise. *The Apollo Spacecraft: A Chronology, Volume I, Through November 7, 1962.* (NASA SP-4009, 1969).

Morse, Mary Louise, and Bays, Jean Kernahan. *The Apollo Spacecraft: A Chronology, Volume II, November 8, 1962-September 30, 1964.* (NASA SP-4009, 1973).

Brooks, Courtney G., and Ertel, Ivan D. *The Apollo Spacecraft: A Chronology, Volume III, October 1, 1964-January 20, 1966.* (NASA SP-4009, 1973).

Ertel, Ivan D., and Newkirk, Roland W., with Brooks, Courtney G. *The Apollo Spacecraft: A Chronology, Volume IV, January 21, 1966-July 13, 1974.* (NASA SP-4009, 1978).

Astronautics and Aeronautics, 1968: Chronology of Science, Technology, and Policy. (NASA SP-4010, 1969).

Newkirk, Roland W., and Ertel, Ivan D., with Brooks, Courtney G. *Skylab: A Chronology.* (NASA SP-4011, 1977).

Van Nimmen, Jane, and Bruno, Leonard C., with Rosholt, Robert L. *NASA Historical Data Book, Vol. I: NASA Resources, 1958-1968.* (NASA SP-4012, 1976, rep. ed. 1988).

Ezell, Linda Neuman. *NASA Historical Data Book, Vol II: Programs and Projects, 1958-1968.* (NASA SP-4012, 1988).

Ezell, Linda Neuman. *NASA Historical Data Book, Vol. III: Programs and Projects, 1969-1978.* (NASA SP-4012, 1988).

Gawdiak, Ihor, with Feder, Helen. *NASA Historical Data Book, Vol. IV: NASA Resources 1969-1978.* (NASA SP-4012, 1994).

Astronautics and Aeronautics, 1969: Chronology of Science, Technology, and Policy. (NASA SP-4014, 1970).

Astronautics and Aeronautics, 1970: Chronology of Science, Technology, and Policy. (NASA SP-4015, 1972).

Astronautics and Aeronautics, 1971: Chronology of Science, Technology, and Policy. (NASA SP-4016, 1972).

Astronautics and Aeronautics, 1972: Chronology of Science, Technology, and Policy. (NASA SP-4017, 1974).

Astronautics and Aeronautics, 1973: Chronology of Science, Technology, and Policy. (NASA SP-4018, 1975).

Astronautics and Aeronautics, 1974: Chronology of Science, Technology, and Policy. (NASA SP-4019, 1977).

Astronautics and Aeronautics, 1975: Chronology of Science, Technology, and Policy. (NASA SP-4020, 1979).

Astronautics and Aeronautics, 1976: Chronology of Science, Technology, and Policy. (NASA SP-4021, 1984).

Astronautics and Aeronautics, 1977: Chronology of Science, Technology, and Policy. (NASA SP-4022, 1986).

Astronautics and Aeronautics, 1978: Chronology of Science, Technology, and Policy. (NASA SP-4023, 1986).

Astronautics and Aeronautics, 1979-1984: Chronology of Science, Technology, and Policy. (NASA SP-4024, 1988).

Astronautics and Aeronautics, 1985: Chronology of Science, Technology, and Policy. (NASA SP-4025, 1990).

Management Histories, NASA SP-4100:

Rosholt, Robert L. *An Administrative History of NASA, 1958-1963.* (NASA SP-4101, 1966).

Levine, Arnold S. *Managing NASA in the Apollo Era.* (NASA SP-4102, 1982).

Roland, Alex. *Model Research: The National Advisory Committee for Aeronautics, 1915-1958.* (NASA SP-4103, 1985).

Fries, Sylvia D. *NASA Engineers and the Age of Apollo* (NASA SP-4104, 1992).

Glennen, T. Keith. *The Birth of NASA: The Diary of T. Keith Glennan,* edited by J.D. Hunley. (NASA SP-4105, 1993).

Project Histories, NASA SP-4200:

Swenson, Loyd S., Jr., Grimwood, James M., and Alexander, Charles C. *This New Ocean: A History of Project Mercury.* (NASA SP-4201, 1966).

Green, Constance McL., and Lomask, Milton. *Vanguard: A History.* (NASA SP-4202, 1970; rep. ed. Smithsonian Institution Press, 1971).

Hacker, Barton C., and Grimwood, James M. *On Shoulders of Titans: A History of Project Gemini.* (NASA SP-4203, 1977).

Benson, Charles D. and Faherty, William Barnaby. *Moonport: A History of Apollo Launch Facilities and Operations.* (NASA SP-4204, 1978).

Brooks, Courtney G., Grimwood, James M., and Swenson, Loyd S., Jr. *Chariots for Apollo: A History of Manned Lunar Spacecraft.* (NASA SP-4205, 1979).

Bilstein, Roger E. *Stages to Saturn: A Technological History of the Apollo/Saturn Launch Vehicles.* (NASA SP-4206, 1980).

Compton, W. David, and Benson, Charles D. *Living and Working in Space: A History of Skylab.* (NASA SP-4208, 1983).

Ezell, Edward Clinton, and Ezell, Linda Neuman. *The Partnership: A History of the Apollo-Soyuz Test Project.* (NASA SP-4209, 1978).

Hall, R. Cargill. *Lunar Impact: A History of Project Ranger.* (NASA SP-4210, 1977).

Newell, Homer E. *Beyond the Atmosphere: Early Years of Space Science.* (NASA SP-4211, 1980).

Ezell, Edward Clinton, and Ezell, Linda Neuman. *On Mars: Exploration of the Red Planet, 1958-1978.* (NASA SP-4212, 1984).

Pitts, John A. *The Human Factor: Biomedicine in the Manned Space Program to 1980.* (NASA SP-4213, 1985).

Compton, W. David. *Where No Man Has Gone Before: A History of Apollo Lunar Exploration Missions.* (NASA SP-4214, 1989).

Naugle, John E. *First Among Equals: The Selection of NASA Space Science Experiments* (NASA SP-4215, 1991).

Wallace, Lane E. *Airborne Trailblazer: Two Decades with NASA Langley's Boeing 737 Flying Laboratory.* (NASA SP-4216, 1994).

Center Histories, NASA SP-4300:

Rosenthal, Alfred. *Venture into Space: Early Years of Goddard Space Flight Center.* (NASA SP-4301, 1985).

Hartman, Edwin, P. *Adventures in Research: A History of Ames Research Center, 1940-1965.* (NASA SP-4302, 1970).

Hallion, Richard P. *On the Frontier: Flight Research at Dryden, 1946-1981.* (NASA SP-4303, 1984).

Muenger, Elizabeth A. *Searching the Horizon: A History of Ames Research Center, 1940-1976.* (NASA SP-4304, 1985).

Hansen, James R. *Engineer in Charge: A History of the Langley Aeronautical Laboratory, 1917-1958.* (NASA SP-4305, 1987).

Dawson, Virginia P. *Engines and Innovation: Lewis Laboratory and American Propulsion Technology.* (NASA SP-4306, 1991).

Dethloff, Henry C. *"Suddenly Tomorrow Came ...": A History of the Johnson Space Center, 1957-1990.* (NASA SP-4307, 1993).

General Histories, NASA SP-4400:

Corliss, William R. *NASA Sounding Rockets, 1958-1968: A Historical Summary.* (NASA SP-4401, 1971).

Wells, Helen T., Whiteley, Susan H., and Karegeannes, Carrie. *Origins of NASA Names.* (NASA SP-4402, 1976).

Anderson, Frank W., Jr., *Orders of Magnitude: A History of NACA and NASA, 1915-1980.* (NASA SP-4403, 1981).

Sloop, John L. *Liquid Hydrogen as a Propulsion Fuel, 1945-1959.* (NASA SP-4404, 1978).

Roland, Alex. *A Spacefaring People: Perspectives on Early Spaceflight.* (NASA SP-4405, 1985).

Bilstein, Roger E. *Orders of Magnitude: A History of the NACA and NASA, 1915-1990.* (NASA SP-4406, 1989).

New Series in NASA History, published by The Johns Hopkins University Press:

Cooper, Henry S. F., Jr. *Before Lift-Off: The Making of a Space Shuttle Crew.* (1987).

McCurdy, Howard E. *The Space Station Decision: Incremental Politics and Technological Choice.* (1990).

Hufbauer, Karl. *Exploring the Sun: Solar Science Since Galileo.* (1991).

McCurdy, Howard E. *Inside NASA: High Technology and Organizational Change in the U.S. Space Program.* (1993).

www.ingramcontent.com/pod-product-compliance
Lightning Source LLC
Chambersburg PA
CBHW081121170526
45165CB00008B/2518